Frontiers in Polymer Sc

(Volume 1)

Industrial Applications of Polymer Composites

Edited by

Subhendu Bhandari

Department of Plastic and Polymer Engineering
Maharashtra Institute of Technology
Aurangabad, India

Prashant Gupta

MIT-Center for Applied Materials Research and Technology
Department of Plastic and Polymer Engineering
Maharashtra Institute of Technology
Aurangabad, India

&

Ayan Dey

Indian Institute of Packaging
Mumbai
India

Frontiers in Polymer Science

(Volume 1)

Industrial Applications of Polymer Composites

Editors: Subhendu Bhandari, Prashant Gupta and Ayan Dey

ISBN (Online): 978-981-5124-81-1

ISBN (Print): 978-981-5124-82-8

ISBN (Paperback): 978-981-5124-83-5

©2023, Bentham Books imprint.

Published by Bentham Science Publishers Pte. Ltd. Singapore. All Rights Reserved.

First published in 2023.

need for a court order if at any point you breach any terms of this License Agreement. In no event will any delay or failure by Bentham Science Publishers in enforcing your compliance with this License Agreement constitute a waiver of any of its rights.

3. You acknowledge that you have read this License Agreement, and agree to be bound by its terms and conditions. To the extent that any other terms and conditions presented on any website of Bentham Science Publishers conflict with, or are inconsistent with, the terms and conditions set out in this License Agreement, you acknowledge that the terms and conditions set out in this License Agreement shall prevail.

Bentham Science Publishers Pte. Ltd.
80 Robinson Road #02-00
Singapore 068898
Singapore
Email: subscriptions@benthamscience.net

BENTHAM SCIENCE

CONTENTS

PREFACE

A polymer composite is a three-dimensional combination of at least an organic or inorganic filler dispersed in a continuous or co-continuous phase of an individual polymer or a polymeric blend. The presence of polymers as well as fillers in different blends may be wisely utilized in different combinations to overcome the limitations of the individual components toward achieving the required characteristics of industrial products. The ability to achieve a set of desired characteristics such as mechanical, chemical, physical, electrical, electrochemical, biological, *etc.*, suiting the needs, processability, dimensional stability, thermal, cost, and so on has allowed polymeric composites to be used in a wide range of industrial applications such as construction, packaging, tissue engineering, energy storage, sensors, transportation, and so on. The consumption of polymer composites for industrial applications is ever-growing with time. With the advent of upgraded and new technologies related to the preparation of individual components and composites, different combinations of materials have attracted researchers from academia as well as industries. To meet the demand of consumers, new material development as well as finding new applications for the existing materials have become the major focus of industrial research. Several books were published on polymer composites in the last decade with a primary focus on materials, characterization, or any specific area of application. However, it is envisaged that a single book encompassing the knowledge related to polymer composites in different fields of application is unavailable in the market. In this book, the focus is on the recent developments in various major sectors where composite materials are very popular and significantly used. With the rich experience in polymer composites and nanocomposites of the editors, especially in application development, technical services, and new product development, we thought of bringing together authors having expertise in polymer composites in specific industrial domains. The outline of the book encompasses relevant knowledge from an application point of view and represents its diversities in a nutshell. Therefore, we feel the proposed book may attract a broad readership from industry as well as academia.

We thank all the contributors for their generous efforts and cooperation in providing chapters highlighting recent research and findings across the globe. We are thankful to all the authors of the studies cited in the present book. We also like to express our gratitude to the entire team of Bentham Science Publishers for their collaboration, prompt assistance, and patience during the publication of this book.

Subhendu Bhandari
Department of Plastic and Polymer Engineering
Maharashtra Institute of Technology
Aurangabad, India

Prashant Gupta
MIT-Center for Applied Materials Research and Technology
Department of Plastic and Polymer Engineering
Maharashtra Institute of Technology
Aurangabad, India

Ayan Dey
Indian Institute of Packaging
Mumbai
India

FOREWORD

This is a great pleasure for me to write a foreword for the book "Industrial Applications of Polymer Composites" which would definitely be a great addition to the library of documents concerning Polymer Science and Technology and more precisely the areas of polymer technology dealing with the applications of different polymer composites, the area of material science now under sharp investigation. Considering the growth of polymer technology to cater to the ever-increasing demand of society, polymer composites have appeared to us as God's blessings. The importance and scope of utilities of polymer composites cannot be overemphasized. The entire world of Material Science has been revolutionized since people could sense the widespread applicability and flexibility of polymer composites which now can be tailor-made. It can possibly be mentioned very concisely that the evolution of material science in the last three decades is the evolution of polymer composites.

The present book is a nice compilation of the scope of utilization of different types of polymer composites. The vision of the editors and the authors who are young and energetic academicians with very good exposure and practical experience in the different fields of construction, packaging, tissue engineering, batteries, microbial fuel cell, sensors and automotive appliances is really praiseworthy. Their tenacious endeavour in presenting the current scenario of the role of polymer composites in fabricating items of the different fields as mentioned above is quite inspiring and interesting. The field of construction has seen a sea change with the advent of polymer composites. The use of multi-layered laminar composites has enabled the civil engineer to substitute a substantial proportion of heavy concrete and thus help to reduce the total weight of the construction, an essential need of time. Many polymeric additives are now available to enhance the flow properties of concrete material. A highly durable, light weight construction is now readily available. The packaging industries have greatly benefitted from the use of polymer composites in a great way. The concept of multi-layered films, each film in its turn being a composite one has enabled the packaging scientists to control permittivity and diffusivity of the various harmful environmental gases and thus prolonging the shelf life of the contents. The synthesis of semi-permeable membrane, the most essential component of microbial fuel cells could not have reached so advanced stage so early without the polymer composites. The different gadgets and accessories meant for the automotive sector would not have been possible to be fabricated without the polymer composites. It is worthwhile mentioning here that the authors having expertise in the respective fields described in the book have tried their best to make the readers acclimatized with the products made up of polymer composites finding applications in different fields for their properties that cannot be challenged by other materials commonly available.

I presume the basic mission of the editors and authors has been successful. The readers would definitely be able to feel and sense the fragrance of the book on reading. I pray to the Almighty for its widespread success.

Debabrata Chakrabarty
Department of Polymer Science & Technology
University of Calcutta
Kolkata, India

List of Contributors

Amandeep Singh	Department of Polymer Science and Technology, University of Calcutta, Kolkata, India
Anusha Vempaty	Department of Life Sciences, School of Basic Sciences and Research, Sharda University, Greater Noida – 201306, India
Arti Rushi	Maharashtra Institute of Technology, Aurangabad, India
Barun Kumar	Department of Life Sciences, School of Basic Sciences and Research, Sharda University, Greater Noida – 201306, India
Bhagwan Ghanshamji Toksha	Maharashtra Institute of Technology, Aurangabad, India
Deepak Jadhav	Department of Agricultural Engineering, Maharashtra Institute of Technology, Aurangabad-431010, India
Dinesh Rathod	Department of Physics, JES, R. G. Bagadia Arts, S. B. Lakhotia Commerce, and R. Bezonji Science College, Jalna, India
Kalpana Sharma	Department of Life Sciences, School of Basic Sciences and Research, Sharda University, Greater Noida – 201306, India
Kunal Datta	Deen Dayal Upadhayay KAUSHAL Kendra, Dr. Babasaheb Ambedkar Marathwada University, Aurangabad (MS), India
Madhuri N. Mangulkar	Department of Civil Engineering, Marathwada Institute of Technology, Aurangabad, India
Naveen Veeramani	Center for Carbon Fiber and Prepregs, CSIR-National Aerospace Laboratories, Bangalore-560017, India
Nabakumar Pramanik	Department of Chemistry, National Institute of Technology, Arunachal Pradesh, Arunachal Pradesh-791113, India
Narayan Chandra Das	Rubber Technology Centre, Indian Institute of Technology, Kharagpur-721302, India
Prosenjit Ghosh	Center for Carbon Fiber and Prepregs, CSIR-National Aerospace Laboratories, Bangalore-560017, India
Rupesh Rohan	Indian Rubber Manufacturers Research Association (IRMRA), Sri City Trade Centre, Sri City, District: Chittoor, Andhara Pradesh, India
Sampad Ghosh	Department of Chemistry, Nalanda College of Engineering, Nalanda-803108, Bihar, India
Shweta Rai	Department of Life Sciences, School of Basic Sciences and Research, Sharda University, Greater Noida – 201306, India
Soumya Pandit	Department of Life Sciences, School of Basic Sciences and Research, Sharda University, Greater Noida-201306, India
Sovan Lal Banerjee	Pritzker School of Molecular Engineering, University of Chicago, Chicago, IL, 60637, USA
Togam Ringu	Department of Chemistry, National Institute of Technology, Arunachal Pradesh, Arunachal Pradesh-791113, India

Tushar Kanti Das   Rubber Technology Centre, Indian Institute of Technology, Kharagpur-721302, India

Vaibhav Raj   Department of Life Sciences, School of Basic Sciences and Research, Sharda University, Greater Noida – 201306, India

CHAPTER 1

Polymer Composites for Construction Applications

Dinesh Rathod[1], **Madhuri N. Mangulkar**[2] and **Bhagwan Ghanshamji Toksha**[3,*]

[1] *Department of Physics, JES, R. G. Bagadia Arts, S. B. Lakhotia Commerce, and R. Bezonji Science College, Jalna, India*

[2] *Department of Civil Engineering, Marathwada Institute of Technology, Aurangabad, India*

[3] *Maharashtra Institute of Technology, Aurangabad, India*

Abstract: Polymer composite concrete (PCC) nowadays plays a major role in the construction industry. PCC is a valuable element in the development of sustainable construction materials. The polymers and classical concrete blends offer newer properties and applications. A polymeric action in the field of admixtures provides insight into the development of highly performing modified mineral concrete and mortars. The influence of various polymers on the properties of concrete is variable due to the polymeric chain reactions. The optimization of properties such as crack resistance, permeability, and durability with the addition of polymer is required. The present work reviews the types, performances, and applications of PCC to improve various properties of concrete in both fresh and hardened states as they have shown a strong potential from technical, economical, and design points of view.

Keywords: Concrete properties, Polymer composite concrete, Polymeric chain.

INTRODUCTION

In the history of mankind, the last two centuries have witnessed rapid advances in construction material technology enabling civil engineers to achieve structures with increased safety and functionality at the economy of scale which could have served the common needs of society [1-4]. The elongation of structural life span against environmental deterioration, sustaining natural calamities such as earthquakes, heavy traffic densities, blast impact from terror attacks, debris flow, and highly corrosive environments demands better quality of reinforced concrete for building structures. The activities encompass the upgradation of design codes and strength requirements. This leads to the exploration of reinforced concrete (RC) materials which could provide strengthening in structures to meet the adequate strength requirements and extend the service life [5-9].

* **Corresponding author Bhagwan Ghanshamji Toksha:** Maharashtra Institute of Technology, Aurangabad, India; E-mail: mittoksha@gmail.com

Subhendu Bhandari, Prashant Gupta and Ayan Dey (Eds.)

The use of polymer additives may be practiced either as a part of a concrete admixture recipe or as external support to already existing structures. The basic composition of concrete is a mixture of fine sand and liquid cement. This formulation is one of the most widely used items in the world after water which is used for construction of all building purposes [10]. The use of concrete is twice as the mixture of aluminium, wood, and plastic as well. As it is widely used globally, it is expected to generate about $600 billion in revenue in the upcoming 5-10 years [11]. Considering the massive use of concrete, there are some disadvantages as they contribute ca. 8% of greenhouse emissions [12-16]. The possible alternatives and related concerns with the usage of supplementary cementitious materials to address the issue of greenhouse gas emissions from the use of concrete were reviewed by Sabbier Miller *et al.* [17]. Other environmental concerns include large-scale illegal sand mining, as well as some effects on the surrounding environment, such as the changes in river surface leaf flow, the effects of urban heat on islands, and the effects of toxic factors on public health. There is a need for extensive research and development to control further damage to the environment still meeting human needs. One of the promising alternatives is to increase the production of secondary raw materials reducing the volume of conventional concrete with still maintaining the construction standards.

The second type of usage of polymer composite as external support includes repair, rehabilitation, and strengthening of structural elements such as beams/columns. The orthodox approaches to rehabilitation, and strengthening structures include the use of an external layer of metallic plate, textile fibre sheet, wire mesh, post tensioning, concrete or steel jacketing, and injection of epoxy [18-22]. The conditions and criteria for selecting one reinforced concrete over another are largely dependent on the type of structure, the degree of strengthening required, and the associated cost. The extent of strengthening and the cost at which it is achieved is a delicate balance to maintain. There are certain challenges with these conventional methods. The heavy weight of externally bonded steel plates requires mechanical fastening and ongoing maintenance to prevent corrosion [18]. The requirement of installation of steel anchorages, deviators, and protection of the steel strand and anchorages against corrosion are some of the downsides of external post-tensioning [24, 25]. These requirements add to the labour and cost of the solution. The need for section enlargement, erection of temporary formwork, and mechanical interlock achieved by the installation of steel dowel bars are un-desirous in concrete jacketing [21, 22]. Polymers and polymer composites were developed very fast compared to other civil engineering tools [23]. Advanced polymer composites have been used the primarily in the aerospace and marine industries, but have also been used in civil engineering for the last few years as they have some unique properties [24]. The scoring points for a particular reinforced concrete system will be fulfilling all design requirements, having the shortest installation time, and realization at the lowest cost including the initial material supply and installation, as well as future maintenance costs such as ongoing corrosion protection, regular inspection, and monitoring. There is a huge amount of polymeric materials available in the form of waste or recycled materials. The adaptation of circular economics modifying the current conventional economy of the construction industry is depicted in Fig (**1**). This model of inclusion of polymer materials is crucial for completing the cycle in a cost-effective way and addressing the environmental issues [25]. Polymer-based materials for reinforced concrete systems are one of the most promising solutions. Their characteristics being non-destructive, light weight, having high tensile strength, corrosion-resistance, lack of long-term maintenance requirements and cost-effectiveness increase their usability in reinforced concrete (RC) materials. Traditional approaches began to take shape in the early 20th century. Concrete-polymer composites including polymer-modified (or cement) mortar and concrete, polymer mortar and concrete, and polymer-impregnated mortar and concrete have been developed in the world for over the past 50 years. In 1965, the first sample of polymer-impregnated concrete (PIC) was discovered at Brookhaven National Laboratory [25]. After this, the whole world was drawn to it [26]. The implementation of polymer-based solutions for reinforced concrete needs to address the issues such as fire performance to provide the necessary fire endurance period without collapse [27]. The other limiting concern with the polymer-based solutions is the limits on the degree of strengthening that may be achieved. The use of fibre polymers in reinforced concrete will improve performances, compared to those realised through other existing techniques [28]. The comparison of stress-strain values of various materials used in concrete composites was recorded by Theodoros Rousakis *et al.* [29].

Polymer concrete is used in specialized construction projects where there is a need to resist several types of corrosion and is supported to have durability *i.e.*, to last longer. It can be used similar to ordinary concrete. Polymer concrete is applied for various construction purposes such as repairing corrosion-damaged concrete [31], pre-stressed concrete [32], nuclear power plants [33], electrical or industrial construction [34, 35], marine works [36], prefabricated structural components like acid tanks, manholes, drains, highway median barriers, waterproofing of structures, sewage works and desalination plants. Contemporary researchers are taking the help of advanced technological tools to explore the field of polymer concrete [37-40]. The use of artificial intelligence has also occupied space in enhancing the field of polymer concrete. The newer models are recently getting evolved for FRP- confined concrete [41]. The comparison between the hybrid models with the existing design relations of the ultimate strain and strain capacities has revealed that the hybrid models have superior abilities in terms of accuracy. The feasibility and accuracy prediction by AI algorithms are utilized for predicting the bond strength between FRPs and concrete [42].

Fig. (1). Inclusion of polymers in construction industry circular economics [30].

POLYMER ADDITIVE FOR ALTERING CLASSICAL CONCRETE

Classically, cement and its mixtures react with other elements to form a hard matrix making the material like durable stone which has many applications [43]. The currently practiced alternatives as a solution to mitigate the pollution of other industries are capturing wastes such as coal fly ash or bauxite tailings and residue with concrete being the main material for a structure that is resilient to weather disasters [44]. Often, additives (such as pozzolans or superplasticizers) are included in the mixture to improve the physical quality of that wet mixer. Concrete is often poured with the reinforcing material which provides higher tensile strength; it is termed reinforced concrete. Many other types of non-cementation concrete are also used, such as asphalt concretes are used for road works, and polymer concretes that are polymer mixtures, are used for building materials and for many new types of works as well [45]. The materials that are used in composite formation encompass carbon, glass, aramid, and basalt fibres that are bonded together by the matrix of a polymer such as epoxy, vinyl ester, or polyester to form various reinforced polymers. The polymer composite concrete for construction is the composite phase made by fully/partially replacing the cement hydrate binders of conventional cement concrete with polymer binders or liquid resins [46]. The hardening of polymer concrete is done *via* liquid resins such as thermosetting resins, methacrylic resins, and tar-modified resins by polymerization at ambient temperature. The binder phase of polymer composite concretes consists of polymers, and does not contain any cement hydrate. The binding of aggregates is done *via* the use of polymeric binders. As compared to the basic ordinary cement concrete, the properties of polymer composite concrete such as strength, adhesion, water-tightness, chemical resistance, freeze-thaw durability, and abrasion resistance are being improved to an extent such that it is considered as a replacement.

The liquid form of polymers (latex) usually is referred to as resin. There are numerous polymeric resins commonly used such as methacrylate, polyester resin, epoxy resin, vinyl ester resin, and furan resins. The use of unsaturated polyester resins is prevalent in resin systems for polymer concrete because of their low cost, easy availability, and good mechanical properties [47]. The downsides of their use are higher flammability and disagreeable odour. However, it has received some attention because of its good workability and low temperature curability [48]. The selection of the particular type of resin depends upon various factors namely the cost, chemical or weather resistance, desired properties, *etc.* For polymer concrete, unsaturated polyester resins are the most versatile polymers used owing to their cost efficiency, good mechanical properties of concrete, and easy availability. The curing effect in case of polymer composite concrete with resins is rapid at ambient temperature. The curing time is reportedly reduced to a greater extent as compared to conventional concrete with polymer concrete developing 80% strength after one day of curing at room temperature while the conventional concrete could gain 20% strength of its 28-day strength in one day [49]. The adhesive nature of polymers brings ease in binding the organic substrate to each other. The selection of a specific resin depends upon factors like cost, desired properties, and chemical/weather resistance required. The choice between epoxy resins and polyester is made by mechanical properties as well as better durability when subjected to harsh environmental factors. The higher cost in case of epoxy resins limits their widespread acceptance. The comparison of epoxy and polymer concrete revealed that traditionally epoxy concrete demonstrated better properties as compared to polyester concrete. The study also concluded that the properties of polyester concrete could be elevated by adding micro fillers and silane coupling agents [50]. The polymer concrete formulations with resin dosage reported in the literature were as high as 20% by weight. The compressive strength of polymer concrete varied as a function of the resin content [51]. Both the compressive strength and flexural strength depend on polymer content. The peak attained would lead to either decreased or unaltered with a further increase in the resin content. The flexural and compressive strength reaching the maximum values between 14 and 16% resin content by weight were reported in the literature. Variation in compressive strength of polymer concrete for various types of resins and their dosage has been reported in the literature [52]. It was observed that the highest strength was obtained in all types of resins at a resin dosage of 12%. For two types of epoxy resins, the strength decreased by increasing the resin content to 15%, whereas, for polyester resin, it almost remained constant. The nature of aggregate used in the polymer concrete formulations was also decisive in deciding optimum resin content. The resin dosage can be set to higher values while using fine aggregate owing to the large surface area of these materials [53, 54].

Polymer composite materials are formed by incorporation of aggregates into polymers [55]. In general, fiber reinforced polymer materials contain high-strength fibres [56]. This material is generally developed using fibers such as natural fibers (jute, kenaf, cellulose), synthetic (glass, carbon, aramid), hybrid fiber (jute/glass, sisal/glass) [57]. Such class of materials is commonly referred to as fiber-reinforced polymer or fiber-reinforced plastic (FRP) composites. Composite materials are applied almost everywhere, because of their unique properties such as binding with another particle, long time workability, durability, crack resistance, *etc.* The sources of composite materials used can be waste materials in some other process. In construction, polymeric composite materials are being used extensively, in which many industrial waste materials can be used as aggregates. One of the most commonly used aggregates for polymeric concrete is glass [58], which can be used in a variety of ways such as, glass fiber, glass dust, and coarse inorganic waste. Also polyester resin and polymer concrete have various aesthetic and structural advantages [59]. Glass fiber reinforced polymeric [60] waste material, incorporated in the polymer matrix, has been used in many construction materials. Additional constituents of the particulates in the polymeric matrix can be in the form of silica, sand, calcium carbonate, mica, white cement, gypsum, perlite, or others. The use of these materials can be cost-effective [61]. The fibers in such compositions have good load-carrying capacity, as well as they are rigid and strong. Moreover, this material has corrosion resistance, good appearance, high temperature resistance, environmental stability, as well as heat, and electrically resistance [62]. There are three types of fiber which are used in the construction industry *i.e.* E-, S- and Z-type glass fiber [63]. Reportedly, carbon fiber is successfully utilized in strengthening and rehabilitation of columns, freeway piers, and chimneys. Consequently, the use of natural fibers can be a great alternative to the sustainable development of the construction sector. However, natural fibers have some drawbacks, such as low durability and low strength. As an alternative to this problem, hybrids of two or more different types of fiber reinforced polymers can be used. *e.g.*, sisal/GFRP, Abaca/jute GFRP, sisal/jute GFRP, and jute GFRP [64]. The types of polymeric materials used in building construction along with their applications, advantages, and disadvantages are presented in Table **1**.

Table 1. Polymeric materials used in building construction along with their applications, advantages and disadvantages.

Polymer Type	Applications	Pros	Cons
Solid Epoxy	Adhesives, flooring, plastics [65].	High strength, bonds to wet or underwater surfaces, excellent chemical resistance, low cure shrinkage.	Fades or yellows over time, high curing time.
Water dispersions or emulsions epoxy	Paints, coatings, primers, sealant [66, 67].	Eco-friendly, easy clean-up, high strength of products, lightweight, inert and non-toxic.	Yellows over time, Limited chemical resistance, high curing time.
Polyurethane	Thermal and acoustical insulation, roofing and sealant [68, 69].	Flexible, tough, abrasion-resistant.	Poor bonding, Moisture caused gas formation, yellows over time.
Polyurea	Protection against moisture, abrasion and corrosion [70, 71].	Flexible, tough, abrasion resistant Fast cure.	Limited bond, Needs primer in critical use, Limited chemical resistance, yellows over time.
Polyester	Anchoring grout, coating, concrete composites to improve the crack resistance and strength [72, 73].	Tough, best chemical resistance, good overlay material.	Shrinks, requirement of primers, styrene odour, toxicity concerns.
Reactive Acrylics	Transparent/translucent sheets, opaque cladding, panel materials, paints [74, 75].	Yellows less, fast cure, Reactive concrete consolidates when epoxy injection is too expensive.	Odour, brittleness, expensive, inhibits curing, low chemical resist.
Acrylic latex	Modify physical and mechanical properties of cement grout [76, 77].	Best cement product modifier, Permeability, Environmental resistance.	Limited freeze-thaw and chemical resistance.
Silicone resin	Making of anti-corrosion, high temperature resistant exterior coatings [78, 79].	Sealant with high service period, color stability, No age hardening.	Expensive, requirement of primer.
Silanes	Flexible and elastic sealing, improved adhesive, high strength concrete [80, 81].	Improved penetrant, Low viscosity, Breathable.	May develop surface cracks.

Synthetic polymer/polymer concrete is one of the alternative options for the Portland cement or construction purpose used to bond a mixture of aggregates together with epoxy resin binders [82]. Polymer concrete is created by different resins and monomers such as epoxy resin [83], polyester [84], and acrylic [85]. The polymer composite concrete thus achieved good resistance against water [86] and it has good durability [87], and good mechanical properties [88]. The adhesive properties of polymer concrete allow the repair of a new kind of cement-based concrete [89]. Epoxy polymer concrete is a type of composite material which is conglutinated aggregate with epoxy resin.

POLYMER COMPOSITE CONCRETE CONCOCTION AND WORKABILITY

The literature reveals different polymer concoctions and their workability in making polymer composite concretes. The research in the direction of usability of polymer concretes aim at elevating their thermal properties such as thermal conductivity, thermo-mechanical properties, and thermo-gravimetric properties. The polymer matrix phases have poor thermal and fire resistance; and along with temperature dependence of mechanical properties that is the hurdle yet to be fully overcome. The experimental parameters such as the glass transition temperature of the polymer matrix phases need to be considered from the viewpoint of thermal properties. The polymer materials have a characteristic behavior by which they generally retain their practical properties at temperatures below the glass transition point and beyond the transition temperatures beginning to decompose thermally. The operational temperature range of the thermoplastic resins may be improved by the addition of suitable cross-linking agents having higher glass transition points. The specific structural requirement with high-voltage electrical insulation is reported by Bowen Xu *et al.* [90]. The liquid styrene and acrylic (SA) monomers, wollastonite, and muscovite mixed in Portland cement led to the polymer composite material. The improvement in dielectric strength as high as 16.5 kV/mm and a reduction in dielectric loss factor by 0.12 along with thermal stability and thermal conductivity were reported.

The epoxy binder is made from resin and amine hardener (BS5462), wherein the epoxy equivalent weight and density were 200 g/equiv. and 1.1 g/cm^3, respectively. Also, the equivalent weight and density of the hydrogen amine and hardener were 100 g/equiv and 0.985 g/cm^3. The granite aggregate was 45 wt. % fine aggregate with a size of less than 4.75 mm, and 55 wt.% coarse aggregates with a size falling between 4.75 mm and 9.5 mm. The weight ratio of 2:1 of epoxy resin was mixed with a hardener and aggregates were incorporated into a mixture. The mixture was finally molded and cured at 25 °C for 72 h [65]. The study reports the use of an epoxy binder with room temperature (RT) viscosity of 5000 and 12000 m.Pa and density in the range 1.42 to 1.48 g/cm^3. The resin/hardener ratio was maintained at 100:60 by weight. The inorganic aggregates used in this are commercially silico-calcaire aggregates, 0/4 sand and 4/10 gravels with actual densities of 2470 and 2530 kg/m^3. The experimental conditions maintained were such that the mixture was dried at 105 degrees for 24 hours before it was ready to use [91]. A control mixer of polymer concrete with two variations of mixtures was set in experimental research. The control mixture of polymer concrete was prepared in a dosage of 12.4% epoxy resin, and a mixture of fly ash and natural river aggregate in a dosage of 12.8%, was prepared in two forms. Sort 1 (0-4 mm) and Sort 2 (4-8 mm) both were in equal proportions. The two waste mixtures were made from epoxy resin, fly ash and 4-8mm sort, and were replaced with only 0-4mm sort dust and chopped PET bottles. The first and second mixtures replaced the dust and PET bottles in 0-4mm sort but kept their proportions at 25%, 50%, 75%, and 100%. To prepare the concrete, fly ash, waste, and PET bottles were mixed and the concrete was prepared by mixing in epoxy resin hardener [92]. The sample was prepared and tested according to the European standardization method [93]. The workability of concrete was reported to increase with increasing PET bottle dosage and decreasing with the increase of waste dosage. It was reported that for resin, diglycidyl ether of bisphenol A (DGEBA) possessing low viscosity was used, and as curing agent polyamine hardener was used. Also, basalt aggregates with a size up to 5 mm were used to study different chemical combinations.

Samples of unfilled and basalt-filled epoxy concrete were prepared using a curing agent DGEBA according to various wt.% of basalt aggregation. The mixture was then poured into a silicone mold and placed in the oven until air bubbles appeared on its surface and then kept at room temperature for a few days. The specimens thus realized were subjected to various tests to test their workability. The workability of this, when the increasing resin into basalt aggregate reaches 25 wt.%, it increased the mechanical properties [94]. Opthophalic polyester resin containing 33wt% styrene as binder, methyl ethyl ketone peroxide (MEKP) as an initiator, and cobalt naphthenate containing 6 wt.% cobalt as a promoter were used in this literature. The particle size of the aggregate was less than 10 mm. Polymer concrete was prepared by mixing crushed gravel, sand, calcium carbonate, and different concentrations of unsaturated polyester resin such as 60, 14, 14 and 12 wt.%. The fresh mixture was poured into the metal mold and allowed to stand for 24 hours at room temperature [95]. The polymers were included in concrete admixtures for improving adhesion to the old surface, flexural strength, tensile strength, and freeze/thaw durability. The mixtures were also useful in reducing permeability, the intrusion of chlorides, salts, and carbon dioxide along with improved abrasion resistance. The polymers replacing natural fine aggregates were reported to reduce chloride ion migration into the concrete samples. The observation is an indication that polymers have inhibited the free chlorides inside concrete [96]. The major concern with the use of a polymer material is the unavailability of all desired functionalities in one material or recipe. Each material brings its own strengths and weaknesses on its inclusion in concrete. Moreover, the effect of polymer addition is not uniform and weak in most of the cases [97]. The polymers are better at ultraviolet light (UV) blocking, and efficient at transmitting water vapor, but are weak at re-emulsification on re-wetting. The condition of meeting the requirement raises the economic point of view. A cost-effective polymer emulsion is required for preparing high-performance concrete [98]. The recycled polymer admixtures have an uneven effect on the concrete, while affecting durability there was no significant effect on mechanical properties [99].

One of the uses of recycled polymer materials could be in asphalt pavements as an alternative to develop sustainable road pavements with increased performance. The use of polymeric waste into asphalt mixture phase prepared by wet and dry processes was reviewed by Duarte and Faxina [100]. The promising use of polymeric waste to improve asphalt pavement performance was echoed in this review with poor storage stability as a concern to overcome. The inclusion of polymer results in an improvement in the durability of asphalt pavement, joint damage, and cracks of cement concrete pavement. The compressive, flexural, tensile strength and impact resistance, flexural fatigue resistance, permeability, frost resistance, shrinkage characteristics and wear resistance of polymer modified cement concrete improved these properties, and the improvement effect was proportional to the polymer content [101].

DURABILITY

Generally, the aspects of durability of a polymer concrete need consideration of humidity, temperature, age of curing, *etc.* The developed polymer concrete material must show higher durability properties for the commercial viability of the product. The basic concrete erodes quickly when exposed to an acidic medium such as sulphuric acid, nitric acid, *etc.* However, the polymer concrete may withstand to such harsh ambient conditions like coming in contact with acidic or alkali medium owing to its high durability [102].The effect of moisture is a critical factor when incorporating the polymer and polyester phases are included in concrete [103]. The curing gets affected substantially *via* water-polyester interaction. The mechanical properties deteriorate with the increase in thermal expansion coefficient as moisture content is increased. The environmental exposure of construction materials is inevitable. The fiber-reinforced polymer is looked as a promising and affordable solution to the corrosion problems of steel reinforcement in structural concrete. The mechanical properties and durability properties of such products are most critical in the applications replacing steel reinforcement. The effect of severe environmental and load conditions for long-time exposures is challenging the polymer replacements. The high pH of the pore water solution, formed during the hydration of the concrete, attacks the chemical behavior of fibres. The use of resin could be involved in the use of glass fibre-reinforced plastic materials. The properties of resin strongly influence the durability of reinforcement reducing the damage caused by environmental cycles [104]. The lignin content in a natural fibre enables better thermal performance whereas the adverse effects of moisture were more aggravated. The ratio of fibre to matrix ratio is limited to the extent of moisture content. There is a significant improvement in the natural fibre resistance towards moisture by the chemical treatments. Thus, there could be possibilities of various blend ratios of chemical additives that need to be employed to achieve a balance between strength and durability requirements for natural fibre composites [105]. The harsh environmental conditions deteriorate the physical and mechanical properties of polymeric composite materials such as strength and flexure modulus [106]. The undesirable degradation was also reported to occur for the adhesive joint in carbon fiber reinforced polymer material. There were enough evidences that clearly depicted weakening of the adhesion between the fibres and matrix over the time interval of 0 to 2000 hrs due to humidity [107].

CONCLUDING REMARKS

The use of polymeric waste materials could solve both the issues of environmental concerns and occupying volumes with increased performance. Among the polymers, the recipes involving acrylic polymers and styrene acrylics demonstrated the best water vapor transmission rates *i.e.*, breathability above all the polymers with styrene acrylics having better water resistance and less UV stability. The requirements of pre-packaged products either in wet or dry conditions could be met by using vinyl acetate ethylene. The styrene-butadiene copolymer resin is very cost-effective with very high adhesive properties. Besides this, the products involving this polymer have better water and abrasion resistance. Though polyvinyl acetate is very cost-effective, hydrolysis of the most re-wettable material in wet alkaline environments leads to an undesirable breakdown of the polymer. There are many different formulations of monomers and each manufacturer combines them to create polymers with specific characteristics. The effects of mixing polymeric materials have uneven effects. Still, there are evidence that the applications of polymer composites could significantly contribute to the construction industry.

ACKNOWLEDGEMENT

Declared none.

REFERENCES

[1]   Van Hoof J, Marston HR, Kazak JK, Buffel T. Ten questions concerning age-friendly cities and communities and the built environment. Build Environ 2021; 199: 107922.
[http://dx.doi.org/10.1016/j.buildenv.2021.107922]

[2]   Manahasa E, Manahasa O. Nostalgia for the lost built environment of a socialist city: An empirical study in post-socialist Tirana. Habitat Int 2022; 119: 102493.
[http://dx.doi.org/10.1016/j.habitatint.2021.102493]

[3]   Tiitu M, Viinikka A, Ojanen M, Saarikoski H. Transcending sectoral boundaries? Discovering built-environment indicators through knowledge co-production for enhanced planning for well-being in Finnish cities. Environ Sci Policy 2021; 126: 177-88.
[http://dx.doi.org/10.1016/j.envsci.2021.09.028]

[4]   Mouratidis K. Urban planning and quality of life: A review of pathways linking the built environment to subjective well-being. Cities 2021; 115: 103229.
[http://dx.doi.org/10.1016/j.cities.2021.103229]

[5]   Yumnam M, Gupta H, Ghosh D, Jaganathan J. Inspection of concrete structures externally reinforced with FRP composites using active infrared thermography: A review. Constr Build Mater 2021; 310: 125265.
[http://dx.doi.org/10.1016/j.conbuildmat.2021.125265]

[6]   Ismail MA, Mueller CT. Minimizing embodied energy of reinforced concrete floor systems in developing countries through shape optimization. Eng Struct 2021; 246: 112955.
[http://dx.doi.org/10.1016/j.engstruct.2021.112955]

[7]   Jayasinghe A, Orr J, Ibell T, Boshoff WP. Minimising embodied carbon in reinforced concrete beams. Eng Struct 2021; 242: 112590.
[http://dx.doi.org/10.1016/j.engstruct.2021.112590]

[8]   Prakash R, Raman SN, Subramanian C, Divyah N. 6 - Eco-friendly fiber-reinforced concretes. In: Colangelo F, Cioffi R, Farina I, Eds. Handbook of sustainable concrete and industrial waste management. Woodhead Publishing 2022; pp. 109-45.
[http://dx.doi.org/10.1016/B978-0-12-821730-6.00031-0]

[9]   Shiping Y, Boxue W, Chenxue Z, Shuang L. Bond performance between textile reinforced concrete (TRC) and brick masonry under conventional environment. Structures 2022; 36: 392-403.
[http://dx.doi.org/10.1016/j.istruc.2021.12.029]

[10]   Gagg CR. Cement and concrete as an engineering material: An historic appraisal and case study analysis. Eng Fail Anal 2014; 40: 114-40.
[http://dx.doi.org/10.1016/j.engfailanal.2014.02.004]

[11]   'Press Releases Archive', Digital Journal. Available From: https://www.digitaljournal.com/pr/ playtreks-officially-launches-treks-a-brand-new-crypto-token-designed-to-complement-the-playtreks-all-in-on-app-for-the-music-industry

[12]   '2018-06-13-making-concrete-change-cement-lehne-preston.pdf'. Accessed: Nov. 09, 2021. [Online]. Available From: https://www.chathamhouse.org/sites/default/files/publications/research/2018-06--3-making-concrete-change-cement-lehne-preston.pdf

[13]   Das S, Saha P, Prajna Jena S, Panda P. Geopolymer concrete: Sustainable green concrete for reduced greenhouse gas emission – A review. Mater Today Proc 2021.
[http://dx.doi.org/10.1016/j.matpr.2021.11.588]

[14] Arrigoni A, Panesar DK, Duhamel M, *et al.* Life cycle greenhouse gas emissions of concrete containing supplementary cementitious materials: cut-off vs. substitution. J Clean Prod 2020; 263: 121465.
[http://dx.doi.org/10.1016/j.jclepro.2020.121465]

[15] Sandanayake M, Lokuge W, Zhang G, Setunge S, Thushar Q. Greenhouse gas emissions during timber and concrete building construction —A scenario based comparative case study. Sustain Cities Soc 2018; 38: 91-7.
[http://dx.doi.org/10.1016/j.scs.2017.12.017]

[16] Fan C, Miller SA. Reducing greenhouse gas emissions for prescribed concrete compressive strength. Constr Build Mater 2018; 167: 918-28.
[http://dx.doi.org/10.1016/j.conbuildmat.2018.02.092]

[17] Miller SA. Supplementary cementitious materials to mitigate greenhouse gas emissions from concrete: can there be too much of a good thing? J Clean Prod 2018; 178: 587-98.
[http://dx.doi.org/10.1016/j.jclepro.2018.01.008]

[18] Li A, Xu S, Wang Y, Wu C, Nie B. Fatigue behavior of corroded steel plates strengthened with CFRP plates. Constr Build Mater 2022; 314: 125707.
[http://dx.doi.org/10.1016/j.conbuildmat.2021.125707]

[19] Lee MJ, Lee K. Performance and cost effectiveness analysis of the active external post tensioning system. J Asian Archit Build Eng 2012; 11(1): 139-46.
[http://dx.doi.org/10.3130/jaabe.11.139]

[20] Schokker AJ. Overcoming challenges in grouting of post tensioned structures. Innovations and developments in concrete materials and construction. Thomas Telford Publishing 2002; pp. 573-8.

[21] Shabani Attar H, Reza Esfahani M, Ramezani A. Experimental investigation of flexural and shear strengthening of RC beams using fiber-reinforced self-consolidating concrete jackets. Structures 2020; 27: 46-53.
[http://dx.doi.org/10.1016/j.istruc.2020.05.032]

[22] Raza S, Khan MKI, Menegon SJ, Tsang HH, Wilson JL. Strengthening and repair of reinforced concrete columns by jacketing: state-of-the-art review. Sustainability (Basel) 2019; 11(11): 3208.
[http://dx.doi.org/10.3390/su11113208]

[23] Bhattacharjee M, James M. Polymers in civil engineering: review of alternative materials for superior performance.2019.
[http://dx.doi.org/10.13140/RG.2.2.22045.67046]

[24] Hollaway LC. The evolution of and the way forward for advanced polymer composites in the civil infrastructure. Constr Build Mater 2003; 17(6-7): 365-78.
[http://dx.doi.org/10.1016/S0950-0618(03)00038-2]

[25] Concrete-Polymer Composites – The Past, Present and Future. Available From: https://www.scientific.net/KEM.466.1 (accessed Nov. 09, 2021).

[26] Frigione M. 16 - Concrete with polymers. In: Pacheco-Torgal F, Jalali S, Labrincha J, John VM, Eds. Eco-Efficient Concrete. Woodhead Publishing 2013; pp. 386-436.
[http://dx.doi.org/10.1533/9780857098993.3.386]

[27] Tretyakov A, Tkalenko I, Wald F. Fire response model of the steel fibre reinforced concrete filled tubular column. J Construct Steel Res 2021; 186: 106884.
[http://dx.doi.org/10.1016/j.jcsr.2021.106884]

[28] Fathelbab FA, Ramadan MS, Al-Tantawy A. Strengthening of RC bridge slabs using CFRP sheets. Alex Eng J 2014; 53(4): 843-54.
[http://dx.doi.org/10.1016/j.aej.2014.09.010]

[29] Rousakis T. Retrofitting and strengthening of contemporary structures: materials used. In: Beer M,

Kougioumtzoglou IA, Patelli E, Au IS-K, Eds. Encyclopedia of Earthquake Engineering. Berlin, Heidelberg: Springer 2021; pp. 1-15.
[http://dx.doi.org/10.1007/978-3-642-36197-5_303-1]

[30]   Alhazmi H, Shah SAR, Anwar MK, Raza A, Ullah MK, Iqbal F. Utilization of polymer concrete composites for a circular economy: a comparative review for assessment of recycling and waste utilization. Polymers (Basel) 2021; 13(13): 2135.
[http://dx.doi.org/10.3390/polym13132135] [PMID: 34209639]

[31]   Pellegrino C, da Porto F, Modena C. Experimental behaviour of reinforced concrete elements repaired with polymer-modified cementicious mortar. Mater Struct 2011; 44(2): 517-27.
[http://dx.doi.org/10.1617/s11527-010-9646-0]

[32]   Jia L, Fang Z, Guadagnini M, Pilakoutas K, Huang Z. Shear behavior of ultra-high-performance concrete beams prestressed with external carbon fiber-reinforced polymer tendons. Front Struct Civ Eng 2021; 15(6): 1426-40.
[http://dx.doi.org/10.1007/s11709-021-0783-z]

[33]   Dauji S, Kulkarni A. Fire resistance and elevated temperature in reinforced concrete members: research needs for india. Journal of The Institution of Engineers (India): Series A 2021; 102(1): 315-33.
[http://dx.doi.org/10.1007/s40030-021-00513-4]

[34]   Mechtcherine V, Wyrzykowski M, Schröfl C, *et al.* Application of super absorbent polymers (SAP) in concrete construction—update of RILEM state-of-the-art report. Mater Struct 2021; 54(2): 80.
[http://dx.doi.org/10.1617/s11527-021-01668-z]

[35]   Barbuta M, Toma IO, Harja M, Toma AM, Gavriloaia C. Behaviour of short polymer-high strength concrete columns under eccentric compression. Arch Civ Mech Eng 2013; 13(1): 119-27.
[http://dx.doi.org/10.1016/j.acme.2012.10.004]

[36]   Sadati S, Moradllo MK, Shekarchi M. Long-term durability of onshore coated concrete —chloride ion and carbonation effects. Front Struct Civ Eng 2016; 10(2): 150-61.
[http://dx.doi.org/10.1007/s11709-016-0341-2]

[37]   Ben Seghier MEA, Kechtegar B, Nait Amar M, Correia JAFO, Trung N-T. Simulation of the ultimate conditions of fibre-reinforced polymer confined concrete using hybrid intelligence models. Eng Fail Anal 2021; 128: 105605.
[http://dx.doi.org/10.1016/j.engfailanal.2021.105605]

[38]   Sonnenschein R, Gajdosova K, Holly I. FRP composites and their using in the construction of bridges. Procedia Eng 2016; 161: 477-82.
[http://dx.doi.org/10.1016/j.proeng.2016.08.665]

[39]   Mugahed Amran YH, Alyousef R, Rashid RSM, Alabduljabbar H, Hung CC. Properties and applications of FRP in strengthening RC structures: A review. Structures 2018; 16: 208-38.
[http://dx.doi.org/10.1016/j.istruc.2018.09.008]

[40]   Derkowski W. Opportunities and risks arising from the properties of frp materials used for structural strengthening. Procedia Eng 2015; 108: 371-9.
[http://dx.doi.org/10.1016/j.proeng.2015.06.160]

[41]   Su M, Zhong Q, Peng H, Li S. Selected machine learning approaches for predicting the interfacial bond strength between FRPs and concrete. Constr Build Mater 2021; 270: 121456.
[http://dx.doi.org/10.1016/j.conbuildmat.2020.121456]

[42]   Fan W, Chen Y, Li J, *et al.* Machine learning applied to the design and inspection of reinforced concrete bridges: Resilient methods and emerging applications. Structures 2021; 33: 3954-63.
[http://dx.doi.org/10.1016/j.istruc.2021.06.110]

[43]   Li Z. Advanced concrete technology. Hoboken, N.J: Wiley 2011.
[http://dx.doi.org/10.1002/9780470950067]

[44]   'Making Concrete Change: Innovation in Low-carbon Cement and Concrete', Chatham House – International Affairs Think Tank, Jun. 13, 2018. Available From: https://www.chathamhouse.org/2018/06/making-concrete-chan-e-innovation-low-carbon-cement-and-concrete (accessed Nov. 09, 2021).

[45]   E. Allen and J. Iano, Fundamentals of building construction: materials and methods. 2013. Accessed: Nov. 10, 2021. [Online]. Available From: http://site.ebrary.com/id/10768991

[46]   Jiang D, Smith DE. Anisotropic mechanical properties of oriented carbon fiber filled polymer composites produced with fused filament fabrication. Addit Manuf 2017; 18: 84-94.
[http://dx.doi.org/10.1016/j.addma.2017.08.006]

[47]   Gorninski JP, Dal Molin DC, Kazmierczak CS. Strength degradation of polymer concrete in acidic environments. Cement Concr Compos 2007; 29(8): 637-45.
[http://dx.doi.org/10.1016/j.cemconcomp.2007.04.001]

[48]   Ohama Y. Recent progress in concrete-polymer composites. Adv Cement Base Mater 1997; 5(2): 31-40.
[http://dx.doi.org/10.1016/S1065-7355(96)00005-3]

[49]   Hong S. Influence of curing conditions on the strength properties of polysulfide polymer concrete. Appl Sci (Basel) 2017; 7(8): 833.
[http://dx.doi.org/10.3390/app7080833]

[50]   Kumar R. A review on epoxy and polyester based polymer concrete and exploration of polyfurfuryl alcohol as polymer concrete. J Polym 2016; 2016: 1-13.
[http://dx.doi.org/10.1155/2016/7249743]

[51]   Carrión F, Montalbán L, Real JI, Real T. Mechanical and physical properties of polyester polymer concrete using recycled aggregates from concrete sleepers. ScientificWorldJournal 2014; 2014: 1-10.
[http://dx.doi.org/10.1155/2014/526346] [PMID: 25243213]

[52]   Huang L, Zhao L, Yan L. Flexural performance of rc beams strengthened with polyester FRP composites. Int J Civ Eng 2018; 16(6): 715-24.
[http://dx.doi.org/10.1007/s40999-016-0140-0]

[53]   Ribeiro MCS, Nóvoa PR, Ferreira AJM, Marques AT. Flexural performance of polyester and epoxy polymer mortars under severe thermal conditions. Cement Concr Compos 2004; 26(7): 803-9.
[http://dx.doi.org/10.1016/S0958-9465(03)00162-8]

[54]   Ferreira AJM, Tavares C, Ribeiro C. Flexural properties of polyester resin concretes. J Poly Eng 2000; 20(6): 459-68.
[http://dx.doi.org/10.1515/POLYENG.2000.20.6.459]

[55]   Phiri J, Gane P, Maloney TC. General overview of graphene: Production, properties and application in polymer composites. Mater Sci Eng B 2017; 215: 9-28.
[http://dx.doi.org/10.1016/j.mseb.2016.10.004]

[56]   Hyde A, He J, Cui X, Lua J, Liu L. Effects of microvoids on strength of unidirectional fiber-reinforced composite materials. Compos, Part B Eng 2020; 187: 107844.
[http://dx.doi.org/10.1016/j.compositesb.2020.107844]

[57]   'Analysis of Mechanical and Thermal Behavior of Sisal Fiber Composites: Review | SpringerLink'. Available From: https://link.springer.com/chapter/10.1007/978-981-13-7643-6_2

[58]   'Tribological and mechanical performance of sisal-filled waste carbon and glass fibre hybrid composites - ScienceDirect'. Available From: https://www.sciencedirect.com/science/article/abs/pii/S1359836817300641 (accessed Nov. 25, 2021).

[59]   'Andreescu et al. et al. - 2017 - The Advantages of High-density Polymer CADCAM Int.pdf'. Accessed: Nov. 25, 2021. [Online]. Available From: https://www.revmaterialeplastice.ro/pdf/7%20ANDREESCU%20F%201%2017.pdf

[60]   'Properties of Normal Concrete, Self-compacting Concrete and Glass Fibre-reinforced Self-compacting Concrete: An Experimental Study - ScienceDirect'. Available From: https://www.sciencedirect.com/science/article/pii/S1877705816345052 (accessed Nov. 11, 2021).

[61]   Sabău E, Udroiu R, Bere P, Buranský I, Miron-Borzan CȘ. A novel polymer concrete composite with gfrp waste: applications, morphology, and porosity characterization. Appl Sci (Basel) 2020; 10(6): 2060.
[http://dx.doi.org/10.3390/app10062060]

[62]   Yuhazri MY, Zulfikar AJ, Ginting A. Fiber reinforced polymer composite as a strengthening of concrete structures: a review. IOP Conf Series Mater Sci Eng 2020; 1003(1): 012135.
[http://dx.doi.org/10.1088/1757-899X/1003/1/012135]

[63]   'Damage analysis of glass fiber reinforced composites - ScienceDirect'. Available From: https://www.sciencedirect.com/science/article/pii/B9780081022900000076

[64]   'A brief review on natural fiber used as a replacement of synthetic fiber in polymer composites | Materials Engineering Research'. Available From: https://www.syncsci.com/journal/MER/article/view/294

[65]   Ma D, Liang Z, Liu Y, *et al.* Mesoscale modeling of epoxy polymer concrete under tension or bending. Compos Struct 2021; 256: 113079.
[http://dx.doi.org/10.1016/j.compstruct.2020.113079]

[66]   Zhu K, Li X, Li J, Wang H, Fei G. Properties and anticorrosion application of acrylic ester/epoxy core–shell emulsions: effects of epoxy value and crosslinking monomer. J Coat Technol Res 2017; 14(6): 1315-24.
[http://dx.doi.org/10.1007/s11998-017-9930-9]

[67]   Jack VL, Ashitkov VA, Mnatsakanov SS, Kalaus EE, Shaltyko LG. Water-based epoxy materials: properties of curing agents and film-forming properties. Mater Struct 1990; 23(6): 442-8.
[http://dx.doi.org/10.1007/BF02472027]

[68]   Sair S, Mansouri S, Tanane O, Abboud Y, El Bouari A. Alfa fiber-polyurethane composite as a thermal and acoustic insulation material for building applications. SN Applied Sciences 2019; 1(7): 667.
[http://dx.doi.org/10.1007/s42452-019-0685-z]

[69]   Guo X, Qian S, Qing Q, Gong J. Reinforcement by polyurethane to stiffness of air-supported fabric formwork for concrete shell construction. J Cent South Univ 2019; 26(9): 2569-77.
[http://dx.doi.org/10.1007/s11771-019-4195-3]

[70]   Wu G, Wang X, Ji C, *et al.* Damage response of polyurea-coated steel plates under combined blast and fragments loading. J Construct Steel Res 2022; 190: 107126.
[http://dx.doi.org/10.1016/j.jcsr.2021.107126]

[71]   Xiao Y, Tang Z, Hong X. Low velocity impact resistance of ceramic/polyurea composite plates: experimental study. J Mech Sci Technol 2021; 35(12): 5425-34.
[http://dx.doi.org/10.1007/s12206-021-1113-z]

[72]   Malagavelli V, Rao Paturu N. Polyester fibers in the concrete an experimental investigation. Adv Mat Res 2011; 261-263: 125-9.
[http://dx.doi.org/10.4028/www.scientific.net/AMR.261-263.125]

[73]   Irfan MH. Polyesters in the construction industry. In: Irfan MH, Ed. Chemistry and technology of thermosetting polymers in construction applications. Dordrecht: Springer Netherlands 1998; pp. 230-9.
[http://dx.doi.org/10.1007/978-94-011-4954-9_9]

[74]   Chan JX, Hassan A, Wong JF, Majeed K. Plastics in outdoor applications. Reference module in materials science and materials engineering. Elsevier 2020.
[http://dx.doi.org/10.1016/B978-0-12-820352-1.00064-X]

[75]	Gupta PA, Bhayani H, Pramanik SK, Rao AC, Deshmukh SP. Cost effective approach of acrylic resin based flooring applications. Constr Build Mater 2015; 79: 48-55.
[http://dx.doi.org/10.1016/j.conbuildmat.2014.12.047]

[76]	Anagnostopoulos CA, Patsios A. Effect of acrylic latex on the properties of cement grouts. Proceedings of the institution of civil engineers - construction materials 2019; 172(3): 144-54.
[http://dx.doi.org/10.1680/jcoma.16.00012]

[77]	Akinyemi B, Omoniyi T. Properties of latex polymer modified mortars reinforced with waste bamboo fibers from construction waste. Buildings 2018; 8(11): 149.
[http://dx.doi.org/10.3390/buildings8110149]

[78]	Lv Y, Wu S, Cui P, *et al.* Environmental and feasible analysis of recycling steel slag as aggregate treated by silicone resin. Constr Build Mater 2021; 299: 123914.
[http://dx.doi.org/10.1016/j.conbuildmat.2021.123914]

[79]	Lork A. Silicone resin emulsion coatings: The strength profile of the silicone resin network in paint microstructures. Surf Coat Int B Coat Trans 2004; 87(1): 41-6.
[http://dx.doi.org/10.1007/BF02699563]

[80]	Xu Q, Zhan S, Xu B, Yang H, Qian X, Ding X. Effect of isobutyl-triethoxy-silane penetrative protective agent on the carbonation resistance of concrete. Journal of Wuhan University of Technology-Mater Sci Ed 2016; 31(1): 139-45.
[http://dx.doi.org/10.1007/s11595-016-1343-6]

[81]	Zander L, Peng J. New silane-terminated polymers for sealants and adhesives. Adhesion adhesives & sealants 2018; 15(2): 20-3.
[http://dx.doi.org/10.1007/s35784-018-0014-8]

[82]	Schmalz P, Griessenauer C, Ogilvy CS, Thomas AJ. Use of an absorbable synthetic polymer dural substitute for repair of dural defects: a technical note. Cureus 2018; 10(1): e2127.
[http://dx.doi.org/10.7759/cureus.2127] [PMID: 29607275]

[83]	Guo SY, Zhang X, Chen J-Z, *et al.* Mechanical and interface bonding properties of epoxy resin reinforced Portland cement repairing mortar. Constr Build Mater 2020; 264: 120715.
[http://dx.doi.org/10.1016/j.conbuildmat.2020.120715]

[84]	Shen Y, Huang J, Ma X, Hao F, Lv J. Experimental study on the free shrinkage of lightweight polymer concrete incorporating waste rubber powder and ceramsite. Compos Struct 2020; 242: 112152.
[http://dx.doi.org/10.1016/j.compstruct.2020.112152]

[85]	Ma D, Liu Y, Zhang N, Jiang Z, Tang L, Xi H. Micromechanical modeling of flexural strength for epoxy polymer concrete. Int J Appl Mech 2017; 9(8): 1750117.
[http://dx.doi.org/10.1142/S1758825117501174]

[86]	'Towards potential applications of cement-polymer composites based on recycled polystyrene foam wastes on construction fields: Impact of exposure to water ecologies - ScienceDirect'. Available From: https://www.sciencedirect.com/science/article/pii/S2214509521001790 (accessed Nov. 25, 2021).

[87]	Frigione M, Lettieri M. Durability issues and challenges for material advancements in frp employed in the construction industry. Polymers (Basel) 2018; 10(3): 247.
[http://dx.doi.org/10.3390/polym10030247] [PMID: 30966282]

[88]	Yeon J. Deformability of bisphenol A-type epoxy resin-based polymer concrete with different hardeners and fillers. Appl Sci (Basel) 2020; 10(4): 1336.
[http://dx.doi.org/10.3390/app10041336]

[89]	He Y, Zhang X, Hooton RD, Zhang X. Effects of interface roughness and interface adhesion on new-to-old concrete bonding. Constr Build Mater 2017; 151: 582-90.
[http://dx.doi.org/10.1016/j.conbuildmat.2017.05.049]

[90]	Xu B, Li H, Bompa DV, Elghazouli AY, Chen J. Performance of polymer cementitious coatings for

high-voltage electrical infrastructure. Infrastructures 2021; 6(9): 125.
[http://dx.doi.org/10.3390/infrastructures6090125]

[91]   Elalaoui O, Ghorbel E, Ouezdou MB. Influence of flame retardant addition on the durability of epoxy based polymer concrete after exposition to elevated temperature. Constr Build Mater 2018; 192: 233-9.
[http://dx.doi.org/10.1016/j.conbuildmat.2018.10.132]

[92]   Sosoi G, Barbuta M, Serbanoiu AA, Babor D, Burlacu A. Wastes as aggregate substitution in polymer concrete. Procedia Manuf 2018; 22: 347-51.
[http://dx.doi.org/10.1016/j.promfg.2018.03.052]

[93]   Shamsuyeva M, Endres HJ. Plastics in the context of the circular economy and sustainable plastics recycling: Comprehensive review on research development, standardization and market. Composites Part C: Open Access 2021; 6: 100168.
[http://dx.doi.org/10.1016/j.jcomc.2021.100168]

[94]   Hassani Niaki M, Fereidoon A, Ghorbanzadeh Ahangari M. Mechanical properties of epoxy/basalt polymer concrete: Experimental and analytical study. Struct Concr 2018; 19(2): 366-73.
[http://dx.doi.org/10.1002/suco.201700003]

[95]   Hashemi MJ, Jamshidi M, Aghdam JH. Investigating fracture mechanics and flexural properties of unsaturated polyester polymer concrete (UP-PC). Constr Build Mater 2018; 163: 767-75.
[http://dx.doi.org/10.1016/j.conbuildmat.2017.12.115]

[96]   Mohammed H, Giuntini F, Sadique M, Shaw A, Bras A. Polymer modified concrete impact on the durability of infrastructure exposed to chloride environments. Constr Build Mater 2022; 317: 125771.
[http://dx.doi.org/10.1016/j.conbuildmat.2021.125771]

[97]   Agostinho LB, Alexandre CP, da Silva EF, Toledo Filho RD. Rheological study of Portland cement pastes modified with superabsorbent polymer and nanosilica. J Build Eng 2021; 34: 102024.
[http://dx.doi.org/10.1016/j.jobe.2020.102024]

[98]   Zhang M, Jing Y, Yang Y, *et al.* The influence of emulsified asphalt on mechanical properties of self-compacting concrete. Constr Build Mater 2021; 297: 123842.
[http://dx.doi.org/10.1016/j.conbuildmat.2021.123842]

[99]   Velardo P, Sáez del Bosque IF, Matías A, Sánchez de Rojas MI, Medina C. Properties of concretes bearing mixed recycled aggregate with polymer-modified surfaces. J Build Eng 2021; 38: 102211.
[http://dx.doi.org/10.1016/j.jobe.2021.102211]

[100]   Duarte GM, Faxina AL. Asphalt concrete mixtures modified with polymeric waste by the wet and dry processes: A literature review. Constr Build Mater 2021; 312: 125408.
[http://dx.doi.org/10.1016/j.conbuildmat.2021.125408]

[101]   Wathiq Hammodat W. Investigate road performance using polymer modified concrete. Mater Today Proc 2021; 42: 2089-94.
[http://dx.doi.org/10.1016/j.matpr.2020.12.290]

[102]   Kiruthika C, Lavanya Prabha S, Neelamegam M. Different aspects of polyester polymer concrete for sustainable construction. Mater Today Proc 2021; 43: 1622-5.
[http://dx.doi.org/10.1016/j.matpr.2020.09.766]

[103]   Haddad H, Al Kobaisi M. Influence of moisture content on the thermal and mechanical properties and curing behavior of polymeric matrix and polymer concrete composite. Mater Des 2013; 49: 850-6.
[http://dx.doi.org/10.1016/j.matdes.2013.01.075]

[104]   Micelli F, Nanni A. Durability of FRP rods for concrete structures. Constr Build Mater 2004; 18(7): 491-503.
[http://dx.doi.org/10.1016/j.conbuildmat.2004.04.012]

[105]   Azwa ZN, Yousif BF, Manalo AC, Karunasena W. A review on the degradability of polymeric composites based on natural fibres. Mater Des 2013; 47: 424-42.

[http://dx.doi.org/10.1016/j.matdes.2012.11.025]

[106] Galvez P, Abenojar J, Martinez MA. Effect of moisture and temperature on the thermal and mechanical properties of a ductile epoxy adhesive for use in steel structures reinforced with CFRP. Compos, Part B Eng 2019; 176: 107194.
[http://dx.doi.org/10.1016/j.compositesb.2019.107194]

[107] Alawsi G, Aldajah S, Rahmaan SA. Impact of humidity on the durability of E-glass/polymer composites. Mater Des 2009; 30(7): 2506-12.
[http://dx.doi.org/10.1016/j.matdes.2008.10.002]

CHAPTER 2

Polymer Composites as Packaging Materials

Amandeep Singh[1,§] and **Sovan Lal Banerjee**[2,*,§]

[1] *Department of Polymer Science and Technology, University of Calcutta, Kolkata, India*

[2] *Pritzker School of Molecular Engineering, University of Chicago, Chicago, IL, 60637, USA*

Abstract: This chapter aims to obtain a better understanding of the role of polymer nanocomposites in different packaging applications such as food packaging, electronic packaging, and industrial packaging. Dispersion of nanoparticles (NPs) in the packaging materials improves the properties like mechanical strength and modulus, water resistance, gas permeability, *etc.* In addition, bioactive agents in the packaging materials impart interesting smart phenomena like antimicrobial, and antifouling properties. Generally, petroleum fuel-based thermoplastic polymers are conventionally used in primary and secondary packaging. Some of the widely used polymeric packaging materials consist of polyethylene terephthalate (PET), high-density polyethylene (HDPE), polyvinyl chloride (PVC), low-density polyethylene (LDPE), polypropylene (PP), and polystyrene (PS). However, as the consequence of the harmful impacts of fossil fuel-based packaging materials on humans, animals, and the environment has become understandable, more and more emphasis has been shifted to biopolymers (cellulose, protein, marine prokaryotes, *etc.*) and their nanocomposites. Bio-based or bio-originated polymers or biopolymers are eco-friendly, non-hazardous to living beings as well as to the environment, biodegradable, abundant, and a better alternative to depletable fossil fuel-based materials. Biopolymer-based nanocomposites advocate all desirable aspects of a packaging material to be sustainable, reliable, and environmentally friendly. In addition, the nature-inspired active and intelligent/smart packaging materials are economical and their contribution to reviving the circular economy is prominent.

Keywords: Biopolymer, Nanocomposite, Polymer, Packaging.

INTRODUCTION

Nowadays, the implementation of nanotechnology in the packaging sector is widely accepted and considered a promising area of research due to its immensely interesting advantages such as gas barrier property, enhanced mechanical strength, increased bioavailability of nutrients, and special features such as antimicrobial,

---

* **Corresponding author Sovan Lal Banerjee:** Pritzker School of Molecular Engineering, University of Chicago, Chicago, IL, 60637, USA; E-mail: sovanbanerjee1987@gmail.com
§ Both the Authors equally contributed to this book chapter

Subhendu Bhandari, Prashant Gupta and Ayan Dey (Eds.)

antifouling, and conductive properties in the packaging materials [1 - 4]. Packaging materials play a significant role in different areas such as food [4], electronics [5], pharma [6], and several other industrial sectors. The contribution of the packaging industry to the overall economy of any nation is significant. For instance, the food packaging industry in the United States contributes approximately $561 billion, and about 15% of this amount goes to the food packaging segment itself. A packaging material provides a shield to the packaged stuff from mechanical hazards during transport or service delivery. Among diverse packaging materials such as glass, paper, metals *etc.*, polymers (thermoplastic, thermosetting, and elastomers) are chosen to be one of the best options as the packaging materials as they have good mechanical strength, manufacturing easiness, and are economical [7]. Petroleum-based polymers and their nanocomposites are being used in packaging for a long time. Polymer nanocomposites (PNCs) are materials where the required polymer/ blend of polymers are reinforced (generally, < 5 wt%) with desired nanoparticles (NPs) having aspect ratios (L/h) of > 300 to impart properties like mechanical strength, gas barrier properties, conductivity, bioactive properties (antimicrobial, antifouling), *etc.* [8]. For the gas barrier property, different kinds of clays *viz.* montmorillonite (MMT), kaolinite (K), layered double hydroxides (LDHs)] having silicate layers have been used, which actually enhance the diffusion path length for any gas/ gasses by forming a tortuous path inside the polymer matrix [9]. Composting with clay nanoparticles also enhances the water vapour resistance (WVR) property of the polymer nanocomposite [10]. As per the International Organization for Standardization [11], composite materials are defined as materials with multiple phases, where two or more materials of different chemical or physical properties are merged together to make a system with one continuous phase and another dispersed phase. Suppose in case of the polymer-clay nanocomposite material, polymer acts as a continuous phase whereas clay plays the role of the dispersed phase.

However, as the harmful effects of the packaging materials derived from fossil-fuel on human, animals, and environment are perspicuous, growing interests in the development of bio-based polymers and their nanocomposites have been observed nowadays [12 - 14]. The biodegradability, easy availability, and non-hazardous nature of the biopolymers to the living being undoubtedly highlight this category of polymer as a better alternative to depletable fossil fuel-based materials. More recently, to judge the environmental impact of biodegradable polymers and their nanocomposites, these have been used in the food industry [14, 15]. Suitable biopolymers for packaging such as cellulose, marine prokaryotes, starch, protein, poly-β-hydroxybutyrate (PHB), and polyhydroxy valerate (PHAs) are mixed with several additives to form the hybrids in order to develop various types of active and intelligent/smart packaging materials. Active food packaging not only supports the transportation of food materials by providing the safest milieu to protect the food, but such packaging also protects from harmful bacteria, contamination, and degradation due to the presence of specific additives in the packaging materials [16]. However, uses of the bio-based polymers are mostly limited to the food packaging industry because of their sustainability, reliability, and environment-friendly nature. But, in the electronic packaging sector, mostly thermosetting polymers are being used like phenolic, epoxy, silicone, and polyester [5]. These polymers serve as an adhesive material to glue a metal led frame to the semiconductor chip. Although, polymer-based electronic packaging material is a cheap alternative compared to the metal and ceramic-based packaging; however, much lower electrical and thermal conductivities are still the issues to be overcome [5]. Recently, the nature-inspired intelligent/smart packaging materials are also paving a new era to the packaging technology [17]. All these applications of the polymer-based nanocomposites in the packaging tors have been detailed in the following sections, which we believe, would be very interesting to the readers as well as to the researchers of this field.

Definition of Composites and Classifications of the Composite Materials

In polymer science, composite materials are defined as the systems consisting of two or more immiscible phases having different physical or chemical properties. There are different methods available to form the nanocomposites such as blending, compounding, filling, melting, mixing, and assembling. These are very well known terms in polymer composite area (Fig. **1**) and can be found elaborately in dedicated literature [18, 19].

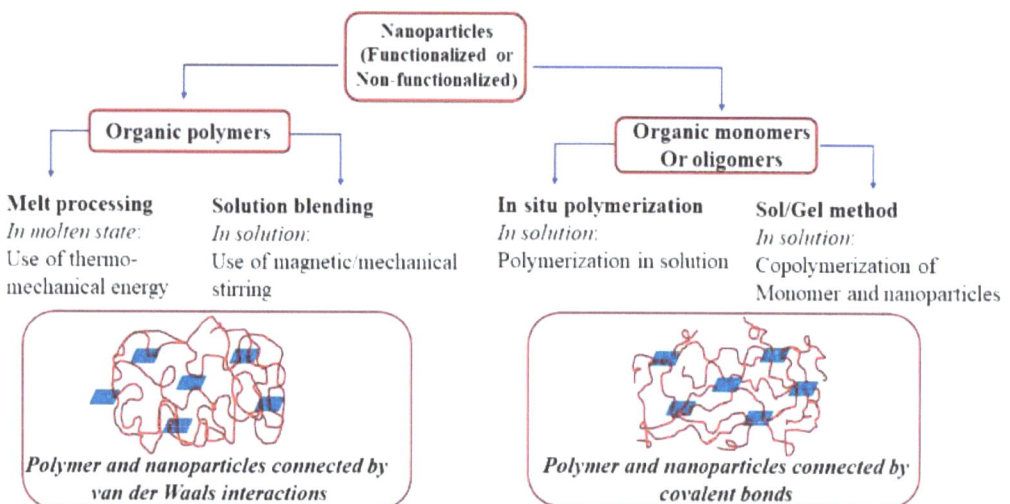

Fig. (1). Outline of polymer nanocomposite preparation techniques [19].

Typically, three types of composites are available like – **(a)** fibrous composite, consisting of reinforcing fibers, **(b)** laminated composite, having multiple layers of materials stacks together using matrix binder, and **(c)** particulate composite, consisting of particles dispersed in the continuous medium. In industries, fillers are used for several purposes such as to increase the mechanical strength of the material, impart several special properties, decrease the linear co-efficient of expansion, improve thermal and electrical conductivities, reduce molding cycles and finally sometimes as a cheapener.

POLYMERS USED IN THE PACKAGING INDUSTRY

Different plastic materials are used in food packaging depending on their specific characteristics and requirements [20, 21]. Some of the widely used polymers in food packaging industry are polyethylene terephthalate (PET), high-density polyvinyl chloride (PVC), polyethylene (HDPE), low-density polyethylene (LDPE), polystyrene (PS), and polypropylene (PP). However, the electronic packaging industries use different resins *i.e.,* epoxy, phenolics *etc.* as a packaging material. The contribution of plastic industry to the total economy of any nation remains significant. The plastic industry of USA is one of the largest industries among the hundreds, adding billions of dollars every year in overall economy. Among a variety of polymer packaging materials, each material is used as per the end-use requirement after making sure of its suitability. For instance, PET is a transparent material and possesses an excellent tensile strength as well as yield strength, but it melts easily on exposure to heat. Thus, it is ideal for cold packaging of beverages [22]. However, HDPE is used in milk beverage packaging because it is comparatively inexpensive in terms of raw materials and processing. It is also used to prepare clouded containers/ bottles, in which a clear/ transparent material is required along with a strong material, for example, milk packaging. Thus, HDPE is suitable where the clarity of the container is not a factor [23]. Generally, PVC is used for wrapping due to its cost efficiency and stretching properties, and also it is easy to extrude in the sheet form [24]. Likewise, LDPE is used for making bags for food storage because of its excellent barrier properties, cost efficiency, and large stretching capacity [25]. PP is used to manufacture comparatively rigid containers such as cups, bowls, baby bottles, *etc.* as it possesses high-strength properties [26]. It is comparatively costlier than other polymers being used for packaging. PS is widely used to manufacture Styrofoam food cups, containers, meat trays, egg trays, *etc.* due to its rigid form and heat resistance properties [27].

Mechanical properties of a material designate how strong it is as well as what type of support, reinforcement, and strength it possesses. The tensile strength of a material usually indicates its resistance towards stress and strain. Barrier property towards various gases and moisture of packaging material is a very important aspect. In many cases, water vapour and oxygen permeability through the packaging material are the major concerns because of water and oxygen sensitivity of the packaged food or substances. However, it is also noticeable that the conventional plastic packaging materials do not address the other problems associated with packaging such as the prevention of contamination, preservation of food, and biodegradability [28, 29].

POLYMER NANOCOMPOSITES FOR PACKAGING

Conventionally, the food-packaging materials were considered to provide merely a barrier shield to the food. However, many advancements have taken placed with time in the field of development of polymer nanocomposite materials for food packaging. Polymer nanocomposites are fabricated by combining polymer/ polymers with different additives using different techniques such as *in-situ* polymerization, solution-mixing process, sol-gel process, melt mixing process, and *in-situ* intercalative polymerization. All these processes have been detailed in the following subsection. Polymer nanocomposite-based packaging materials possess additional properties compared to packaging materials prepared from an individual polymer. In addition, the focus is being shifted from the conventional polymeric materials, those were based on fossil resources, to the bio-based polymers, named as biopolymers. If a monomer is obtained from a natural source and the polymer developed from such monomers is present in nature or may be artificially synthesized, the produced polymer is regarded as a biopolymer or a natural polymer [30]. In short, biopolymers are nature-based macromolecules originating from animals, plants, or microorganisms. Broadly, biopolymers are of three categories. First, such biomaterials that are directly taken from the natural raw materials (*i.e.*, cellulose, starch, marine prokaryotes, protein); second, biomaterials obtained from a chemical conversion from the bio-based precursors; and third, biomaterials that are acquired from various microorganisms (*i.e.*, poly-β-hydroxybutyrate (PHB), polyhydroxyvalerate (PHAs)).

METHODS OF PREPARING POLYMER NANOCOMPOSITES

For fabrication of the nanocomposites, different kinds of nanoparticles such as zero-dimensional (nanodots), one-dimensional (clay platelets), two-dimensional (nanofibers, nanotubes), three-dimensional (spherical metal oxide nanoparticles), and some amorphous materials have been used as per the requirements [11]. In the case of the layered material/ polymer nanocomposites, the formation of a nanocomposite involves three basic steps, namely **(a)** intercalation, **(b)** flocculation, and **(c)** exfoliation [11]. Depending on the adopted method of preparation, the formed nanocomposite material can adopt any one or a mixture of these entire phenomena. It is obvious that the exfoliated nanocomposites always have better properties compared to the intercalated or flocculated structures [31]. The most used methods for the fabrication of polymer nanocomposites have been explained below:

Melt Mixing/ Melt Intercalation

The melt mixing process involves the dispersion of the layered, spherical, or amorphous nanoparticles in the polymer matrix by means of mechanical shearing in the molten state of the polymer. Injection molding, and extrusion molding techniques adopt this traditional method to prepare the polymer nanocomposites, which are also the mostly used processes by the industries [31].

In-situ Intercalative Polymerization Process

The *in-situ* intercalative polymerization process was the very first method utilized to prepare a polymer/clay nanocomposite (Nylon 6/ clay nanocomposite). In this method, the layered silicate is used to be swollen in the liquid monomer which undergoes polymerization in the presence of an initiator and stimuli such as heat or ultraviolet (UV) light. This process is very much effective to fabricate exfoliated polymer nanocomposites [32].

Solution Mixing Process

The solution mixing process is also a well-adopted process to prepare the polymer nanocomposite. In this method, nanofillers (*e.g.*, nanolayered silicates) are dispersed and swollen in a solvent, and in the same solvent, the polymer or prepolymer is also solubilized. Gradually, the polymer/prepolymer replaces the entrapped solvent inside the silicate layer and the composite is formed subsequently [33].

ADDITIVES FOR POLYMERS

Additive materials are the chemical compounds added into a polymer matrix to increase or impart some specific property as per the requirement of the end-user. The basic polymer matrix is combined with a formulary having different additives during shaping of the polymer, through extrusion, vacuum molding, injection molding, blow molding, *etc.* to increase the functionality, performance, and aging properties of the final product. The most frequently used additives in various types of polymer packaging materials are plasticizers, antioxidants, acid scavengers, flame retardants, pigments, lubricants, light/ heat stabilizers, slip compounds, thermal stabilizers, antistatic agents, *etc.* [34, 35]. Each additive plays a separate role in enhancing the desired properties of the final product. All the above-mentioned additives can be broadly categorized into the following four classes:

  i. **Functional additives**. These include stabilizers, antistatic agents, plasticizers, slip agents, flame-retardants, lubricants, foaming agents, curing agents, and biocides.
 ii. **Colorants**. These include various pigments and soluble azo colorants.
iii. **Fillers**. These consist of small particle sized powdery substances such as talc, mica, clay, kaolin, barium sulphate, calcium carbonate, *etc.*
 iv. **Reinforcements**. These consist of glass fibers, carbon fibers, *etc.*

However, many countries have implemented legislation to encourage the use of biodegradable additives in polyolefin as well as PET. The basic goal behind the adaptation of biodegradable additives during the processing of conventionally used plastics is the assimilation of those additives back to the environment, which makes the process greener. Some most frequently used additives to be incorporated into the polymer matrix to obtain a polymer nanocomposite material for food packaging applications are discussed below:

Plasticizers

Plasticizers are extensively used in polymer processing because they decrease the rigidity of the 3-dimensional (3D) polymer structure *via* reducing the polymer-polymer interactions. This process helps to improve the workability as well as the durability of the polymer by improving the deformability without any rupture. Chemically, plasticizers are the relatively non-volatile chemical compounds of low molecular weights, which are conventionally used in plastic goods production including the food packaging and epoxy resin coating on cans to be used for foods and beverage storage [36]. For the first time, phthalic acid esters or phthalates are used as the plasticizers. The most frequently used phthalates are diisononyl phthalate (DINP), diisodecyl phthalate (DIDP), and di(2-ethylhexyl) phthalate (DEHP). However, nowadays, uses of phthalates are regulated and even restricted in many products due to their toxicity to human health as well as to the environment [37]. As we have mentioned early, those biopolymers have the potential to replace synthetic polymers in food packaging applications, but because of the brittleness and fragility of the thermo-treated biopolymers, their utilization in the packaging industries is still limited. Therefore, involvement of suitable plasticizers can resolve these issues related to biopolymers by enhancing the processability, flexibility, and durability of products.

Antioxidant

Antioxidants are added into different polymer matrices to be used for packaging to hinder the total oxidative degradation of the material if somehow, during the delivery or transportation, the packaging material gets exposed to the ultraviolet (UV) light. The extremely reactive free radicals are formed on exposure to heat, radiation, and mechanical shear, which can cause degradation of the polymeric material. As far as food packaging is concerned, the possibility of oxidation increases if the packaged food is exposed to high temperatures, it may be by either contact with hot foods, exposure to infrared heating, retort processing, or microwave heating. The most used antioxidants are arylamines followed by the organophosphates (used to decrease hydroperoxides produced while oxidation of alcohols) and phenolics in polymer food packaging materials.

Stabilizers

Generally, in food packaging, epoxy stabilizers are incorporated into polymers. These are the derivatives of epoxidized soybean oil, linseed oil, and sunflower oil. There are even more effective stabilizers available, but they are not considered appropriate for use in food packaging materials due to their potential toxicity, thus are restricted or not recommended.

Fillers

Metal or Metal Oxide Nanoparticles

Nanotechnology in packaging focuses on four different aspects: **(a)** to advance the packaging stability, **(b)** to ameliorate barrier properties by controlling gas exchange, **(c)** to add special features such as antimicrobial and antifouling functionality, and **(d)** to develop intelligent/smart packaging for auto communicating with buyer/users [38]. Commonly used metal nanoparticles include indium (III) oxide (In_2O_3), titanium dioxide (TiO_2), tin (IV) oxide (SnO_2), zinc oxide (ZnO), and silicon dioxide (SiO_2). Presently, in the European Union, only limited nanomaterials such as titanium nitride, zinc oxide, silica, and carbon black are permitted to be used in food packaging materials [39]. However, In Asia and the US, various nanoclays and silver nanoparticles are also acceptable for use in the development of polymer nanocomposites for food packaging. Similarly, the inert nanoscale fillers, for instances, silica nanoparticles, titanium dioxide, titanium nitride particles, chitin/chitosan *etc.* are amalgamated with polymer matrices and are also used as a coating material to ameliorate the stability of packaging materials by increasing its stiffness and mechanical properties [40]. A perfect amalgamation of material science and nanotechnology is a suitable method to design functional nanocomposite materials and their implementations in smart packaging applications [41]. In active packaging, silver nanoparticles (Ag NPs) are the mostly used metal nanoparticles in polymer matrices to impart antimicrobial property to the packaging materials [42].

Clays

In 1990, nanoclays were first utilized in the food packaging application to enhance the mechanical as well as water and gas barrier properties of the packaging materials [43]. Nanoclays have plenty of distinguishable structural properties such as high aspect ratio, flaky and soft platelet-like structure along with the thickness in nanoscale range, low specific gravity *etc.* [44]. There are different forms of nanoclays used for food packaging applications such as organophilic montmorillonite (MMT). For example, organic modified MMT (OMMT), pristine MMT and MMT-Na^+ may be used in food packaging because of their excellent compatibility with mostly used thermoplastic polymers, high specific surface area as well as large aspect ratio in the range 50 to 1000 [45, 46]. In nature, MMT clays are found in bentonite clay; and because of their hydrophilic nature, these are only seemed to be compatible with polymers such as PVA, PLA *etc.* However, in conventional packaging, polymers like PET, PP, HDPE, LDPE *etc.* are used which belong to petroleum-based hydrophobic polymers. Due to increase of the compatibility with these polymers, MMT clays (usually contain Na^+, K^+, or Ca^{2+}) are surface-modified with organic cations conventionally by incorporation of quaternary ammonium salts using ion-exchange chemical reactions. It is observed that OMMT has comparatively lower surface energy and shows higher affinity for Petro-originated polymers [47]. Interestingly, OMMT has higher basal spacing due to the bulkiness and the repulsions between quaternary ammonium ions, which facilitates better insertion of polymer chains into the layered structures of clays [48]. In addition, a starch-based bio-nanocomposite having clay nanoparticles was found to be a green food packaging material due to its high mechanical properties along with biodegradability [49]. Classifications of the nanomaterials have been shown in Fig. (**2**).

TYPES OF POLYMER NANOCOMPOSITES PACKAGING

Active Packaging

Active food packaging products can be made up of both conventional plastics as well as biopolymer-based nanocomposite materials. However, special emphasis has been paid on the biopolymers and related nanocomposites due to their special characteristics such as abundance (*e.g.*, chitin is the 2nd abundant biopolymer in the world), eco-friendliness, renewable sources, non-hazardousness towards nature and organisms *etc.* Active packaging can be defined as the special method to incorporate some extra features in the conventional packaging materials to do some dedicated chemical and physical phenomena to safely retain the quality of the packaged food. This consists of the introduction of some special substances which can absorb carbon dioxide, oxygen, ethylene, moisture, flavours/odours, and in some cases can behave as aroma emitters, antioxidants, antimicrobial agents, sensor to water vapour and oxygen permission *etc.* Among different techniques, moisture absorbers, oxygen scavengers, and antimicrobial packaging are encompassing more than 80% of the recent market. These sensory properties of the active packaging allow to remove undesirable tastes and flavours, and retains the colour or smell of the packaged food which eventually improves the quality and shelf-life of the food [50]. All these properties directly influence the process of reducing the package-related hazardous waste and improve the environment pollution-related issues out of the plastic packaging products [51].

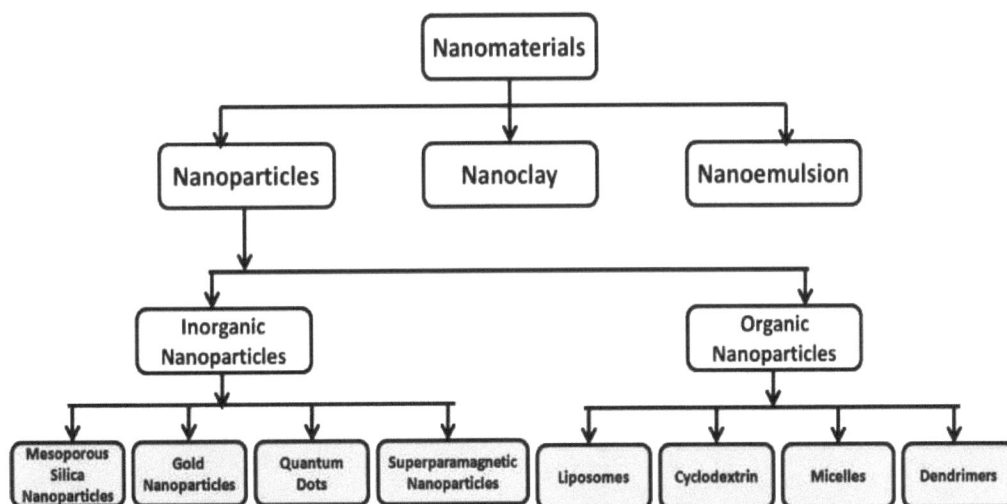

Fig. (2). Classification of the nanomaterials.

The most well-known features of active packaging materials are discussed below:

Antimicrobial Property

Nowadays, antimicrobial food packaging is an interesting topic because of the consumer's awareness. Consumers always prefer a food with the least preservatives. The use of antimicrobial chemical compounds in the packaging material is a distinct approach compared to the direct addition of preservatives in foods [52, 53]. The mechanism of antimicrobial action of silver nanoparticles has been shown in Fig. (**3**) [53]. Interestingly, the rate of antimicrobial diffusion from packaging material to food can be tuned/controlled, whereas the preservatives are supposed to be consumed along with food. It means that the food consumed quickly will have less antimicrobial exposure, and the food taken at the end of its shelf-life will release more antimicrobials from the packaging material [54]. The advantage of this phenomenon is that tuneable inherent glass-transition temperature (an amorphous to crystalline state transition) of polymeric packaging material regulates the release of antimicrobial into food resulting in an effective method of increasing the shelf-life. Therefore, the selection of an appropriate polymer with specific glass transition temperature is very important and selective in this antimicrobial packaging purpose.

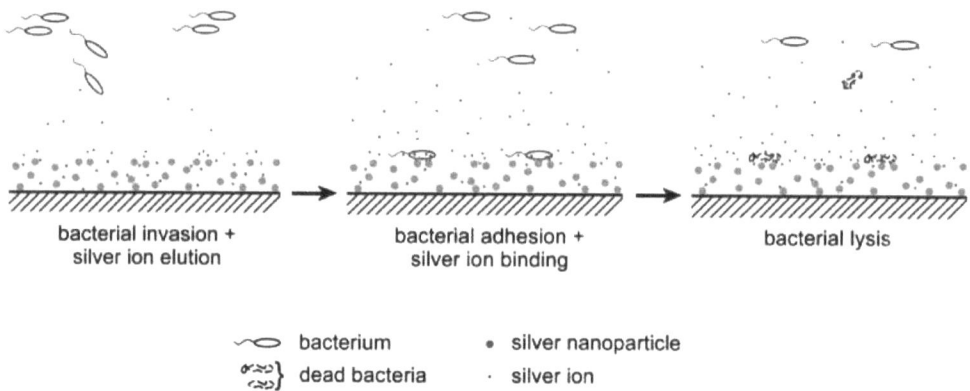

Fig. (3). Antibacterial mechanism of silver. A surface coating containing silver nanoparticles will slowly release silver ions into the coating layer and subsequently the solution. Silver ions will bind the bacterial membrane and proteins, causing cell lysis. The silver ions can originate from the solution, but may also be transferred directly from the surface exposed silver to the bacteria without being dissolved in the medium [53].

Biobased nanocomposite films for food packaging were developed with OMMT clay manifested with a good antimicrobial functionality against *Gram-positive* and *Gram-negative* bacteria [55] due to the presence of quaternary ammonium groups in the organically modified clay. Fabrication of a poly (ε-caprolactone) (PCL) film incorporated with different types of nanoclay against *Staphylococcus aureus* and *Escherichia coli* showed significantly strong antimicrobial potentiality [56, 57]. In different studies, antimicrobial functionality of MMT-Na$^+$ and OMMT was juxtaposed using three different surface modifiers. As a result, the PVOH nanocomposite films fabricated with MMT-Na$^+$, Nanocor® I.44PSS, and Nanocor® I.24TL showed no antimicrobial functionality, like a pristine PVOH film, whereas the nanocomposite film incorporated with Nanocor® I.34TCN exposed substantial antimicrobial potentiality towards *Listeria monocytogenes* and *S. aureus* because of the presence of the quaternary ammonium group [58].

Similar result was found in other research, where a PLA film was incorporated with Cloisite®20A, Cloisite® 30B, and MMT-Na⁺. It was found that the Cloisite® 30B possess more antimicrobial potentiality towards *L. Monocytogenes* compared to MMT-Na⁺ and Cloisite® 20A [59].

Antimicrobial nanocomposite food packaging is one of the most prophesised smart packaging systems that increase the shelf-life of food being sealed by reducing or constraining the food spoiling pathogenic microorganisms [60]. Polymer nanocomposite-based food packaging material having antimicrobial functionality is strikingly beneficial because of the high surface-to-volume ratio of nano fillers. Comparing the bulk counterpart to the improved surface reactivity of nano-sized antimicrobial agents, it helps to inactivate/kill microorganisms [61]. Blending of antimicrobial materials to polymeric matrix is advanced in antimicrobial packaging film during the polymer processing. Thus, polymer nanocomposite materials have been explored for antimicrobial activity such as growth inhibitor, antimicrobial agents, antimicrobial carriers, or antimicrobial food packaging films. On the other hand, metal and metal oxide nanoparticles are widely used nanomaterials for antimicrobial food packaging applications as they exhibit intense antimicrobial activity due to their large surface area and high specificity. The antibacterial properties of metal and its oxide nanoparticles, such as silver, copper, titanium, and zinc nanoparticles have attracted much more attention in food packaging [62]. Moreover, the nanocomposites of silver nanoparticles (Ag NPs) and different biopolymers like chitosan, gelatine, and agar [63] show strong antimicrobial activity against *Gram-positive* and *Gram-negative* bacteria. Copper nanoparticles, owing to profound antimicrobial activity, has attracted the attention to researchers in the application of food packaging. Comparing those two above metal nanoparticles, copper-based nanomaterials are superior to the silver counterparts due to cost effectiveness, insignificant sensitivity to human tissues, and high sensitivity to microorganisms. In another research [60], nanocomposite of agar polymer and six different types of nanoparticles were developed and showed high antimicrobial activity against *Gram-positive* and *Gram-negative* food-borne pathogens. Metal oxides such as TiO_2, ZnO, and MgO are also used for developing antimicrobial packaging films due to their effective antimicrobial activity with high stability [64].

Oxygen Scavenging Property

In packaged foods, the presence of oxygen results in many unnecessary reactions such as, colour changes, nutrient losses, microbial growth, and off-flavour development. It also predominantly affects the ethylene production and respiration rate in vegetables and fruits. However, the passive method of barrier packaging cannot remove oxygen whether O_2-sensitive food has been packed using passive barrier packaging materials such as high-barrier packaging materials with multi-layered structures containing ethylene vinyl alcohol copolymers or aluminium foil [65] or barrier nanocomposites [66]. Inside the container wall, the oxygen may remain in the headspace or dissolve in the food, or permeate the package. To solve such difficulty, an active packaging method using oxygen scavenger system is designed to decrease the residual oxygen in package. However, the risk of anaerobic pathogenic bacterial growth remains also high. Conventionally, oxygen scavenger is packed and sealed in small sachets which are inserted into the package or fixed by adhering to the inner wall of the package materials. Although this technology is well established, there are some issues associated with this method like the accidental consumption of the sachets and the problem in the recycling of such sachets. Polymer nanocomposite could be used as an alternate approach to solve such problems [67]. Aegis® OX is one of the commercially prepared resins with oxygen-scavenging potential, which is a mixture of the active barrier (nylon) and passive oxygen scavengers (nanocomposite clay particles) to enhance the barrier functionality against O_2, CO_2, and aroma. In a report, different edible and biodegradable biopolymer films of whey protein and ascorbic acid with oxygen scavenging performance were manufactured [68]. In another study, various oxygen scavenger films were fabricated by blending different polymers with TiO_2 nanoparticles. Thus, the nanocomposite films with oxygen scavenger performance could be used as active packaging smart material for oxygen-sensitive food products [69].

Ethylene Scavenging Property

Ethylene is a plant hormone (also called ripening hormone) that has physiological effects on fresh fruit and vegetables. It quickens the respiration rate following in maturity and senescence as well as softening and ripening of fruits. In addition, various post-harvest disorders in fresh fruits and vegetables occur because of the excess ethylene accumulation causing yellowish or greenish change in the fruit or vegetables. The accretion of ethylene in the packaged food should be avoided to increase the shelf-life and maintain the quality of packaged food. Various ethylene-absorbing substances are reported, but the efficiency of the materials is difficult to substantiate because of inadequate documentation. It has been exploited for the removal of ethylene vapour which delays the ripening of climacteric fruits due to the photocatalytic effect of TiO_2. As TiO_2 is not being consumed in the reaction, it has unlimited ethylene scavenging capability unlike other ethylene scavengers. In a study, TiO_2 embedded polypropylene films were prepared for ethylene gas removal of packaged horticultural products [70]. Researchers have observed and compared the efficacy of micro (~5 μm) versus nano (~7 nm) TiO_2 particles and found the higher ethylene-scavenging efficacy with nano TiO_2.

Carbon Dioxide Absorbing or Emitting Property

A high concentration of CO_2 level (10–80%) is required for meat and poultry products to prevent the surface microbial growth and increase the shelf-life of product. Removal of oxygen from the package produces a partial vacuum, which results in the collapse of flexible packaging. Therefore, besides this removal process, the simultaneous release of carbon dioxide from inserted sachets, which also consume oxygen, is preferable. Ferrous carbonate or a mixture of sodium bicarbonate and ascorbic acid can be used to develop such systems. In carbon dioxide sachets, calcium hydroxide, potassium hydroxide, sodium hydroxide, calcium oxide, and silica gel can be used in order to remove carbon dioxide at the time of storage and to prevent bursting of the package [29]. Calcium hydroxide is the most common CO_2 scavenger, which reacts with CO_2 and forms calcium carbonate in the presence of high moisture. However, it has a disadvantage that calcium hydroxide irreversibly scavenges the CO_2 from the package headspace resulting in the depletion of CO_2, which is not always desirable [71].

Nano Coating for Various Purposes

Coatings are important layers formed on the base of the packaging materials. Plastic films ornamented with metals like aluminium have been used as gas barriers, light barriers, and as decorative films. By vacuum deposition techniques, the aluminium layer is converted and laid down into typically a few nanometres thick. To prevent corrosion, scratching, and abrasion, causing spoilage of optical properties of the food packaging, the metal layer is sandwiched between different multilayers film constructions. Metallic oxide nanoparticles such as TiO_2, MgO, ZnO, and Al_2O_3, and metallic nanoparticles such as Ag are extensively used to design and development of nanocoating on polymeric films, metallic surface, or paperboard [72]. Nanocoating materials have some novel properties such as optical, mechanical, chemical, electronic, magnetic, and thermal features, which are properly used in some industries including the packaging industry. A wide range of production and precipitation methods of nano-thin films or nanocoatings are industrially known, which include physical vapour deposition, chemical vapour deposition, electronic precipitation/electronic coating, sol–gel process, electrodeposition, rotating coating, spray coating, and self-assembling [73]. Hybrid organic–inorganic nanocomposite coatings of high barrier property could be fabricated by the sol–gel process [74] and are being constructed for oxygen-diffusion barriers for plastics such as PET. This type of nanocoating is fabricated through atmospheric plasma technology using dielectric barrier discharges. The coatings have been reported to be very effective at keeping out oxygen and reserving carbon dioxide, and may be a substitute of the traditional active packaging technologies such as oxygen scavengers. Nanocoatings of materials are used to generate corrosion-resistant, scratch-resistant, antireflective or antimicrobial surfaces. The scratch and abrasion resistance of coatings was increased by nanoscale silicate and alumina without interfering with the transparencies [75]. In a study, using an ultrasonic irradiation method ZnO-coated glass was made and it exhibited a significant antibacterial potential against *Gram-positive* and *Gram-negative* bacteria [76]. The glass slide was coated with a low level of ZnO coating (as low as 0.13%, mean diameter of ZnO nanocrystals of 300 nm). TiO_2 orientated polypropylene based films showed good antibacterial activity against *Escherichia coli* and reduced the microbial contamination on the surface of cut lettuce, reducing the risk level of microbial growth [77]. Bio-hybrid nanocomposite (chitosan and bentonite nanoclay) ornamented on argon-plasm- -activated LDPE-coated paper had improved barrier properties against water vapour, oxygen, grease, and ultraviolet (UV)-light transmission [78]. This coating material of packaging was categorized as 'generally recognized as safe' (GRAS) and the total migration was measured in the allowed range ($\leq$ 6 mg/dm^2) of legislation. The multilayer-coated films were marked as safe and environmentally sound alternatives for synthetic barrier packaging materials. In non-cytotoxic coating for methacrylic thermosets, the antimicrobial activity of silver nanoparticles has been utilized by using lactose-modified chitosan and Ag NPs [79]. Such antimicrobial Ag NPs coated biocompatible polymeric films should have potentiality for antimicrobial active packaging material. Self-cleaning smart nanocoatings that destroy bacteria, isolate pathogens, or fluoresce under certain conditions are under development.

Controlled-Release Kinetics

A substantial amount of the current research activity is dedicated to the encapsulation application apart from using nanoclay as reinforcement. In a study [80], use of OMMT as an active carrier was developed for antibacterial packaging and controlled release study was analysed properly. By melt interaction process, the carvacrol OMMT was dispersed in LDPE film. The antibacterial activity against *Escherichia coli* and *Listeria monocytogenes* was performed along with the storage time. The film of LDPE/ carvacrol totally loses their efficacy within the first month from production date, while LDPE/carvacrol OMMT films retains their efficacy up to a year. Additionally, from the kaolin group, halloysite nanoclays exhibited control and release of active ingredients. Its tubular structure varies from 500 to 1000 nm in length and 15 to 100 nm in inner diameter [81]. Other researchers [82] established the use of active halloysite nanoclay for active packaging. Halloysite surface was modified with cucurbit [6]uril (CB [6]) molecules for enriching the affinity of the nanoclay towards peppermint essential oil before dispersing in pectin solution. Favourable response in antimicrobial and antioxidant efficacies were shown by the functionalized pectin bionanocomposite films with a food simulant (50 v/v% ethanol). Another study [83] revealed that during the film production through extrusion method, the addition of a surfactant in fixing the volatile essential oil with nanoclay structure could minimize the loss of active compounds against degradation and evaporation. As a result, significant antioxidation and antimicrobial activity of the LDPE nanocomposite films are shown against foodborne pathogenic bacteria.

Edible Packaging Materials

In recent time, the use of biodegradable materials, which are derived from the food ingredients such as proteins, polysaccharides, and lipids are found to be a promising replacement of the traditional plastics for the packaging industries. Due to their biodegradability, these materials can be used as edible packaging films or coatings with the direct contact of the packaged food [84, 85]. Wet or dry manufacturing processes are the most convenient techniques to design free-standing edible/ biodegradable packaging films that could be used between the food components in a container or as a wrapping of food stuffs. In addition, edible coatings can be directly applied on the surface of the food products using several techniques such as spraying, dipping, and panning. It is a bit conflicting whether an edible packaging can really be consumable either fully or partially, but this kind of packaging can obviously be used as a delivery vector for many active ingredients such as antioxidants, antimicrobial agents, flavourings *etc.* [86]. Edible coatings and films may also be used to hinder the migration of carbon dioxide, oxygen, and moisture to improve the shelf-life of the food. Proteins and polysaccharides-based hydrocolloids are widely used components for the manufacturing of edible films and coatings [86, 87]. Preparation of the edible films or coatings mostly involves one film forming macromolecules, a solvent, and a plasticizer to provide the flexibility to the film. Schematic of the core properties expected from the intelligent polymer nanocomposite-based packaging materials have been shown in Fig. (**4**).

SMART/ INTELLIGENT PACKAGING

Smart packaging is the ability of the packaging material to monitor quality of the packaged food, which includes the detection of temperature changes or existence of any undesired chemicals using different sensors. These kinds of smart packages can communicate any change in situation of the food by using indicators or sensors [29, 88]. A sensor is a device that can be used to detect or quantify matter or energy, and subsequently inform the users about any physical or chemical change occurred. The sensors conventionally consist of two functional units: a receptor, which is used to transform physical or chemical information to a special form of energy, and a transducer, is responsible for the transformation of that energy to an analytical signal. One of the first examples of intelligent packaging materials are based on the titanium dioxide (TiO_2) nanoparticles embedded sensors that was able to sense the existence of gasses or a change in the pH value by a colour change. A change in the colour of the packet from colourless to blue is observed in the presence of oxygen and this implies that the packet had been opened [40]. Additionally, temperature is also an important factor to maintain the freshness of food. A combination of gold-silver nanoparticles can be used to create a temperature sensor that could change its colour from red to yellow to green depending on the interruption in the cold chain. Even, nano silver can also be used as a sensor to indicate the freshness of the packed meat. The working principle involves the reaction of sulphide, which is the decay product of the spoiled meat with nano silvers to form black coloured silver sulphide [89 - 91]. Although smart packaging has huge potential, all these above-mentioned products are solely allowed in the USA, but not in EU.

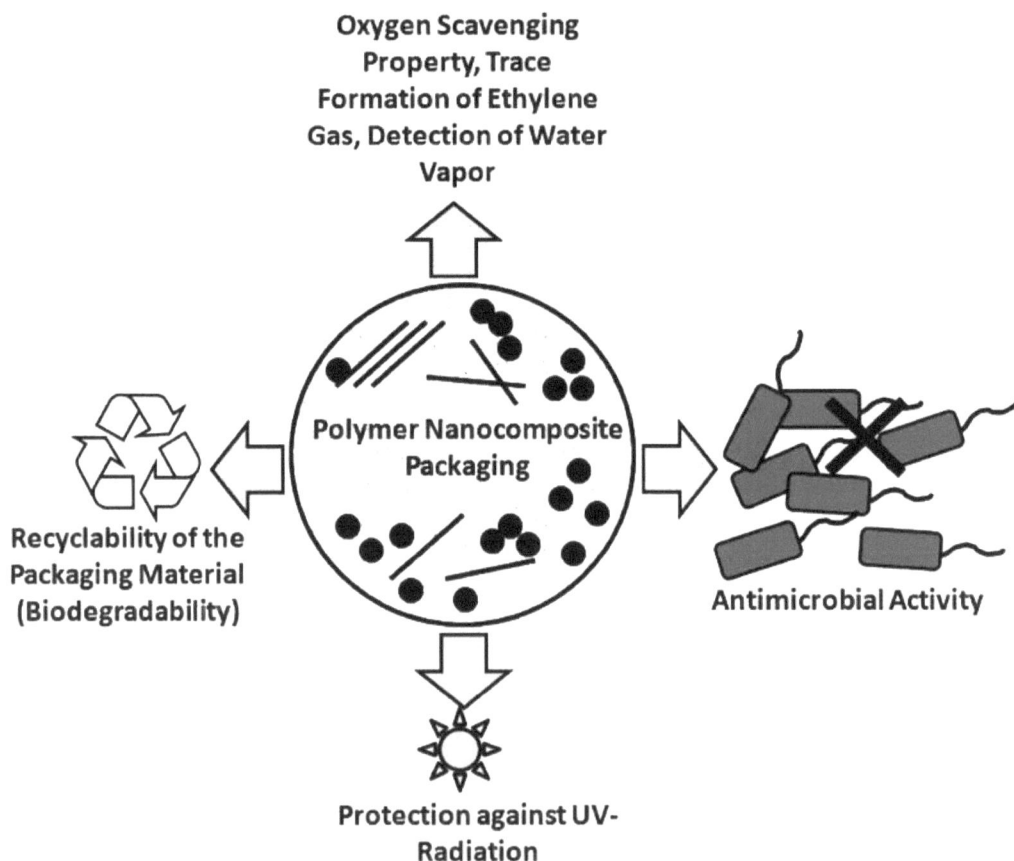

Fig. (4). Some of the core properties of the polymer nanocomposite-based packaging material.

All intelligent or smart packaging materials consist of either an internal or external indicators, which are used to monitor the quality of the packaged food and its surrounding environment to make it safe for the users [88, 91]. By doing so, an intelligent packaging assures the product safety, quality (including generation of harmful chemicals, pathogens, and toxins), and the integrity of the packaging. Moreover, smart packaging can also assure the product authenticity, and in some cases, this kind of packaging can be used to trace a product. Intelligent packaging devices comprise of different important properties such as gas sensing dyes, time-temperature indicators, sensors, microbial growth indicators, physical shock indicators, anti-counterfeiting technologies, tamper proof and anti-theft technologies *etc.* Numerous bioactive nanocomposite materials such as nanostructure indicators, nano sensors, DNA-based biochips, and antigen-detecting biosensors are fabricated using antibodies, high activity of enzymes, microorganisms, or some physicochemical reactions [89]. Due to various interesting properties of silver and gold nanoparticles, magnetic nanoparticles, carbon nanotubes, and quantum dots, several biosensors have been developed based on the physical, chemical, magnetic, optical, electrochemical properties of the mentioned nanomaterials [92, 93]. Biosensors can be used to make polymer films or smart packaging materials to identify allergens, pesticides, pathogens, any temperature changes, leakage, gas contaminations *etc.* [92, 93]. The use of bio-analytical sensors offers numerous advantages such as rapid and high-throughput detection, reduced energy consumptions, cost-effectiveness, and easier recycling [93].

Gas Indicators

Food is a complex material, which can undergo respiration and may change the internal atmosphere while in the packaged state. Gas indicators are especially helpful to detect or monitor any change in the composition of gasses due to the interactions of food with the internal atmosphere of a packet/container by changing its colour *via* chemical or enzymatic reactions [94, 95]. The indicators should be in direct contact with the gaseous environment that surrounds the packaged food as well as the indicators must be capable of sensing any gas leakage in the package. Although most of the gas indicators are designed to detect oxygen and carbon dioxide, they are also being developed to sense ethanol, water vapour, and hydrogen sulphide [90]. Presence of oxygen in food atmosphere can generate unwanted colour changes of the packaged food, oxidative rancidity, and allows the growth of aerobic microbes on foods. Oxygen indicators typically show the colour change in the presence of oxygen, which also designates that the package has been tampered or it has a leak, or the package is not sealed properly.

Temperature Indicator

There are different kinds of inks available in the market that can change their colours depending on temperature variation. These inks can be included in the intelligent packaging so that a clear massage could be conveyed to the consumers about the thermal status of the product. Thermochromic inks are gradually becoming popular to the beverage industries as these indicators can visibly identify the difference between a hot and cold container [96 - 98]. The limitation of thermochromic inks is that they can be significantly affected by temperatures more than 121 °C and in the presence of UV light [98].

Time-Temperature Indicator (TTI)

Time-temperature indicators (TTI) are simple and inexpensive gadgets, which are attached to the package. Commonly, three types of TTI are being recognized: critical temperature indicators that indicate temperature status of a product, which has been heated above or cooled below a predetermined temperature. The second type of TTI is the partial history indicators, which show any sudden change in temperature that can hamper the product quality. Finally, a full temperature history indicator can show a complete temperature profile of the food from the supply chain from the distributor to consumers. The working mechanism of TTIs is very simple. This system detects the time and temperature-dependant changes in the electrochemical, enzymatic, chemical, mechanical, or microbiological changes of the food products [99]. The changes are mostly colorimetric changes, however, mechanical deformations are also a possible way to convey the message [100] to the user, which makes TTIs a very user-friendly device [101]. Fresh-check from Lifeline technologies is an example of a TTI indicator, which involves a colour change based on the polymerization reaction [102].

Freshness Indicators

Freshness indicators are used to directly indicate the freshness of the packaged food. During the growth of the microorganisms in the product, it forms metabolites, which eventually react with the indicator to generate a signal of microbiological quality. It is a major concern for the food industry due to the acute risk of the consumer's health posed by mold, bacteria, and yeast [103]. Mostly, freshness indicators change their colour in contact with the metabolites that have been formed by the microorganisms to give a clear signal to the users about the freshness of the packaged food [104].

Colorimetric Indicator System

Recently, a smart packaging material containing nanoclay based colorimetric indicator has been reported by Gutierrez *et al.* [105]. A blueberry extract was impregnated between the interlayer of nanoclay platelets. The modified MMT and OMMT were able to sense any change in the pH for both alkaline and acid mediums. Polymer nanocomposite films made from this modified OMMT clay could be a potential colorimetric indicator system for fishery products, meat, and fresh food. Pirsa *et al.* [106] have reported a biodegradable colorimetric indicator nanocomposite film consisting of starch-clay hybrid and its use as a milk spoilage indicator. Addition of nanoclay in the composition of bottle reduces the solubility of starch in water and hinders leaching of dye into the milk.

Radio Frequency Identification (RFID)

Radio frequency identification (RFID) tags are believed to be the future of smart food packaging [107]. RFID tags are somewhat advanced data information transporters that can detect and trace a packaged product. Currently, RFID tags are mostly used in expensive products and livestock [108]. This system contains a microchip, which is connected to a tiny antenna. The as mentioned set up allows the RFID tags to be read even in a range of 100 feet [109]. Like the conventional bar-code, RFID does not need to be in a direct line of sight of a scanner. This will allow reading many RFID tags at a rapid rate. In addition, RFID tags could also store information such as relative humidity data, temperature, cooking instructions, nutritional information, *etc.* RFID tags could be used with other sensors such as time temperature indicators, biosensors *etc.*

Microwave Doneness Indicators (MDIs)

Procedures involved in the microwave meals are the anticipation of microwave doneness indicators (MDIs) [7, 110]. These indicators can sense the readiness of the food, which are under microwave cooking and let the consumers know about the status. However, the biggest limitation in this field is the inability to evenly distribute heat in the food during microwave and confirm whether the food is ready to eat or not. Uneven distribution of heat could form the hot spot in the food and trigger the MDIs even though the food is undercooked. An MDI is usually placed either on the lid or the dome of the microwave container so that the signal shown by the indicator is clearly visible to the users. However, the uniform heat distribution from the bottom to the lid of the container is the prime concern of this process to facilitate that the MDIs do not provide false information to the consumer. Apart from this, the container under microwave process should be visible clearly to the consumer so that there is no need to stop the microwave process again and again. As of now, MDIs are not commercially available, but this technique should surely be an important smart sensing system in the near future.

Biosensors for Pathogen or Toxin Identification

For the food industry, food borne pathogens are of prime concern and with time, consumers are aware about this fact. Rapid detection of the trace amount of toxins or pathogens in food is a crucial step to keep the consumer safe from any health hazards. The operating mechanism of a biosensor, which is an analytical device, is simple. Biosensor detects any substance of concern; here, which is pathogen, and transmits the signal to a device to make it quantifiable. In an interesting work, antibodies were attached to the plastic packaging surface to identify toxins or pathogens [111]. During the sensing process, packaging material displayed a visual cue whenever antibodies met the targeted pathogens to alert the consumer. The limitations of these intelligent packaging materials are that they could only sense the toxins or pathogens at their very high concentrations, and they are limited to sense the pathogens that exist at the surface of food. Whereas, a small number of pathogens or toxins are sufficient to make a consumer sick, which conveys a false sense of security to the users of this intelligent packaging. Therefore, this system still needs many modifications before it appears commercially available.

MODIFIED ATMOSPHERE PACKAGING (MAP)

Modified atmosphere packaging (MAP) technique is mostly applied for the packaging of perishable products such as meat, poultry, milk products, vegetables, bakery products, cooked products *etc.*, which are very much prone to be affected by the anaerobic pathogens like *Clostridium botulinum* [112]. MAP technique is a special type of packaging technique where the food is packaged in the presence of a modified atmosphere except air, which is suitable for the long shelf-life of the product [113]. Gases used for MAP technique, are oxygen, nitrogen, and carbon dioxide. In this kind of smart packaging, the designed atmosphere inside the package is able to alter depending on the state of the packaged material (more accurately, respiration of the packaged food). This eventually improves the working-life of the packaged food.

VACUUM PACKAGING

Vacuum packaging is a well-known and an old technique to be used in the smart packaging of varieties of products particularly those that are packaged in cans, jars, pouches, and trays. For this technique, a special device has been used to generate vacuum inside the package [114]. Vacuum packaging especially hinders the growth of aerobic microorganisms like mold over the surface of the food and in due course prevents the degradation of the packaged food. Additionally, this kind of packaging inhibits the aerial oxidation of the packaged foods, which improves the loss of certain vitamins and actual colour from the food. Interestingly, vacuum packaging reduces the volume of the packaging and provides rigidity to flexible packaging.

POLYMER NANOCOMPOSITES IN PACKAGING APPLICATIONS

Polymer Nanocomposites for Food Packaging

In food packaging industry, polymer nanocomposites play an important role in different aspects. Different parameters like price, safety, recyclability, size of packaging, standardization *etc.* are overcome by various inventory steps using technologies including barrier films, active packaging, nanotechnology, microperforation, plasma edible films and coatings with bio-active substances, and far-infrared (FIR) treatments [11, 115]. Recently, different types of packaging processes are employed for improving the quality and shelf-life of packaging meat products. During storage, environmental conditions and stability are checked until the consumption of the products [116].

For processing and packaging, the products safety issue is one of the concerning factors [117]. Quality of packaging food is subjected to some factors such as the type of materials used, use of additives and method of manufacturing. In case of vacuum and MAP (modified atmosphere packaging), plastic films are used and this is effective to some variables such as moisture barriers, sealing features, shrinking factors, cook-in and retort capability [11].

Micro-perforation is one of the new innovative processes regarding packaging materials. It monitors the gas and water vapor permeation and FIR treatments in nanotechnology. Incorporation of bio-agents onto the surface of packaging materials (active packaging), controlling gas and water vapor permeability (passive packaging) quality improvement and shelf-life factors have been considered significantly [118].

Antimicrobial packaging systems impart a major role in food packaging recently. These include different methods like adding a sachet into the package, (O_2 scavenging technology, CO_2 generators, and chlorine oxide generators), dispersing bioactive agents in packaging (bacteriocins, spices, oils, enzymes, preservatives, and additives), *etc.* Antimicrobial macromolecules are also used with film forming properties or edible matrices [59, 119, 120].

Recently, the bio-based natural packaging materials like plant extracts, bacteriocins have been focused due to the consumer's demand and biodegradability, recyclability factors and incorporation of antimicrobial materials into the packaging films minimize the risk factor of food borne organisms [52].

Natural biopolymer-layered silicate composites demonstrate improved packaging and properties due to the size dispersion (in nm). The use of appropriate packaging materials, safety measurements and wholesome products have always been the focus of attention in food packaging. Many types of methods and technologies have been studied to offer good quality, safe foods and to limit environmental pollution. Active packaging (antimicrobial packaging) is the new approach, which has a potential application in packaging of different range of products (fish, meat, poultry, bread, cheese, fruit and vegetables) [121]. Antimicrobial activities in these films prevent the growth of spoilage and pathogenic microorganisms [121].

Polymer Nanocomposites for Application in Fresh Juice Packaging

In recent times, the demand for natural orange juice with minimal or no thermal treatment has raised remarkably because of the concern regarding the nutritional values and physicochemical properties of the packaged juice. Orange juice, which is one of the globally known food products, has very short shelf-life under refrigeration due to the enhanced microbial spoilage. In 2010, Ag and ZnO nanoparticles filled antimicrobial LDPE (0.25% and 1% nano ZnO and 1.5 and 5% for P105) nanocomposite packaging material (film) was developed by Emamifar *et al.* [122] and it was used to preserve and increase the shelf-life of orange juice. The as-prepared packages were instantly wrapped in aluminium foil and subsequently sanitized for 20 min at 95°C. Packages containing orange juices were stored in dark and cold condition (4°C). After that, their microbial and sensory properties were evaluated after 7, 28, and 56 days respectively. In their continuing study, Emamifar *et al.* used Ag and ZnO containing nanocomposite-packaging film and studied its antimicrobial effect over *Lactobacillus Plantrum* in orange juice. Souza *et al.* observed that a lower storage temperature of the unpasteurized orange juice has higher sensory recognition compared to the temperature of 72 h [123].

Polymer Nanocomposites for Application in the Printable or Flexible Electronic Packaging

Polymer nanocomposites have an important role in expanding advanced printing technology. This is comparatively a new approach than the other packaging techniques and requires more characterization and optimization of practical application. In recent days, significant progress has been made with some modification in the semiconductor packaging technology utilizing different printing methods such as ink-jet printing, screen-printing, micro contact printing, *etc.* A totally additive-based non-contacting deposition process is utilized in this method, and it offers a preferable or flexible production. So, this technique is in demand for large area, low cost, flexible and light-weight devices. The electronic application of printable high performance nanocomposite materials requires adhesives (both conductive and non-conductive), interlayer dielectric (low k, low loss dielectrics), embedded passive circuits (capacitors, resistors), *etc.* due to the additive nature of the printing method. Printable materials should be chemically and physically inert to other functions like dielectric, photo-imageable materials processing and in other electrical devices or packages. They must have long operational life also. Some key factors like inductors, an embedded laser and optical interconnects and some printable optically/magnetically active nanocomposites or polymeric materials are also investigated and prepared for the design and fabrication of devices.

In the large area of construction of mechanical flexible devices, organic/polymer, and nanocomposite materials [124, 125] have drawn keen interest. Many substrates and structural modification of nanocomposite materials are widely used because of good compatibility and processing suitability. Several materials such as polyimide, PTFE and liquid crystal polymer (LCP) have been used in this fabrication. Haroun and Youssef studied different characteristics and electrical conductivity of novel poly (methyl methacrylate) (PMMA)-based composites [126]. Paper-based [127] or natural cellulosic fibers and different conductive polymers were also developed by using unbleached bagasse pulp with a conducting polymer (polyaniline). In antistatic packaging material for electronic apparatus and in anti-bacterial paper packaging, this new composite shows a potential application.

Industrial Applications of Polymer Nanocomposites

Apart from the paper, metal, and glass-based packaging products, polymer nanocomposite-based packaging goods are generously used nowadays, which were limited few years back due to the large production cost. Some of the leading companies in this sector are Mitsubishi Gas and Chemical, Honeywell, Triton Systems, Bayer, and Nanocor. Combination of paper and polymer nanocomposite-based packaging products is widely used in many packaging sectors especially in food packaging. Nassar *et al.* [128] reported the utilization of recycled carton paper as an antimicrobial packaging material after coating with Ag NPs impregnated polystyrene (PS). Rice straw powder was utilized as a green reducing and stabilizing agent for the formed Ag NPs. SIG Chromoplasts P Company has produced nanosilica-coated PET bottles, which are utilized for packaging of carbonated water. It was observed that the shelf-life of the product was significantly enhanced three times (25 weeks).

Along with the spherical metal/metal oxide nanoparticles, layered silicates are also used extensively in different packaging sectors especially in beverage packaging including packaging of carbonated water, beer *etc.* Currently, clay NPs have acquired 70% of nanocomposite packaging market volume. Bayer polymers recently created a low cost nanoclay/ polymer composite coating for the interior coating of paperboard carton to enhance the shelf life of packaged juice. Nanoclay reinforced PET bottles have been invented by Nanocor Company, which are distributed by Colour Matrix. This composite using PET bottle was found to be very beneficial to increase the shelf-life of the packaged beer from 11 weeks to 30 weeks.

Apart from this, specially designed nanomaterials and their composites are used for the smart/ intelligent packaging such as the detection of pathogens, tracking of the product tampering, spoilage of the food, *etc.* For example, fluorescent dye modified bacteria antibodies have been used as biosensors in smart packaging. This is found to be very much effective to rapid detection of pathogens like *Salmonella* spp., *Staphylococcus* enterotoxin B, *Listeria monocytogenes*, and *E. coli*- [129]. California's Oxonica Company fabricated nanobarcodes comprising a combination of Ag and Au NPs in different concentrations to detect different products.

CHALLENGES

However, past reviews have recently surfaced many issues with the packaging polymers currently used in the industry. These issues can lead to the leaching of toxic chemicals, the environmental impact on packaging polymers, and economic impact of the industry.

Health Issue: Material-Food Contact

Different chemical reactions form macromolecules from plastic materials corresponding to the respective monomers. From packaging materials, both monomers and oligomers tend to convert into foods. When the concentrations of unreacted monomers or low-molecular-weight substances in food reach a certain limit, health risk arises subsequently, and the human body could strongly be affected by this effect. For instance, residual styrene from PS food packaging can migrate and may result in health issues. It has been reported that epoxy resins of BPA, also known as bisphenol A diglyceride ether (BADGE), give some cytotoxic effects in living tissues and may increase the rate of cell division. Although, recent FDA (food and drug administration) studies, in collaboration with the National Center for Toxicological Research (NCTR), have observed the safety measurement and use of BPA in containers and other food-packaging materials. Some material-food interactions in the current food packaging plastics have been discovered which cause additives within materials to leak into the food. This can be toxic in many cases, causing harm to whoever consumes the food as well as allowing the food to become contaminated with bacteria. The barrier and mechanical properties are lost due to leaching, and some toxicity occurs if food is consumed in the toxic conditions like contamination of bacteria [130]. As previously discussed, integration of additives into the matrix of a polymer alters the molecular structure of the plastic and allows also the incorporation of additive itself within the material [29]. Distortion or loosening effect may be witnessed if the polymer is heated or exposed to UV radiation [131]. This can cause some of the additives to be released and "leak" out of the matrix bond that was holding it [132]. Many nontoxic additives are generally inert; but, upon binding to the polymer matrix, they can interact with food and become harmful in large doses. As the fats and oils are easily solvable in the well of the additive, it might be toxic if consumed. Another study showed the leaching of PVC chemicals and additives with a variation in temperature. It was found that in PVC polymers (greater than 100°C), chemicals would leach from the plastic as this study is not involved in the testing of food specifically. Another study found the toxicity of PS cups as additives leaching into water along with styrene particles [133]. The group found in the styrene chemicals and additives, would leach from the cup into the water in unsafe amounts, when hot water is poured into the styrene cups. Reuse of plastic containers has also been managed by the studies. Schmid's group observed that when PET bottles are reused and sanitized using solar water disinfection and exposed to UV for 6-9 hours while filled with water, the plastic leaches the additives [134]. They also found that leaching increased if the bottles were exposed to UV light and heated to 60°C. Though the additives may help obtain some mechanical and barrier properties that are necessary for food packaging, they can also be hazardous if they interact with food and leach harmful chemicals. In common food packaging of polymers, microwave and heating process are dangerous as toxins can leach from the plastics into food, and may be very harmful to human health. Therefore, there is a requirement to solve the obstacles and then to find a method for measuring additive leaching. However, it is difficult to measure leaching and the various values for diffusivity that are calculated [135]. Studies have noted different ranges of diffusivity in the order of two times of magnitude in LDPE [136] and ten times in PET [137]. Each study still found additive leaching which could be harmful to anyone consuming toxins although results may vary thoroughly.

Pollution Issue: Disposal of Polymer Packaging Materials

We all know that many food packaging plastics are not biodegradable, which gradually increase the waste plastic load, leading to the harmful effects on the environment. Billions of tons of plastic packaging materials are used as landfills under the plastic recycling programs. Although there is no such standard method available to optimize plastic packaging materials' volume control, a reduction in the plastic usage or use of biodegradable polymers in the packaging sector may help to reduce the overwhelming issue [138]. In the United States alone, a humongous 14.5 million tons of plastic waste was generated back in 2018 [139]. An enormous amount of 5,764 million pounds of PET jars and bottles were gathered for recycling in 2013 as per the National Association for PET Container Resources (NAPCOR). Approximately, 85% of those bottles (ca. 4,899 million pounds) were used for food and beverages. As per the NAPCOR report, in 2013, about 369 million pounds of recycled PET bottles were further used to prepare new packaging bottles for food and beverages. Still a huge amount of plastic waste was dumped as landfills, which too affects the environment due to leaching of the toxic compounds [140]. Therefore, it is important to find alternate solutions to manage the huge packaging waste that is generated each year. Reuses of waste plastics are required to meet specific government standards costing as well as sophisticated instruments. Lack of nationwide consciousness about plastic recycling and availability of limited numbers of recycling units significantly make this process an inconvenient one to the everyday consumers [141, 142]. Therefore, considerable changes are required in plastic recycling programs as well as a good amount of subsidy should be included in the nationwide plastic recycling cost to make this more popular to common people and effectively reduce the amount of plastics in landfills.

Commercial Issue: Cost of Polymer Packaging Material

In a year, approximately $84 billion was spent by the food industry for the packaging and processing purposes. Among which, a total of 8% of the cost was spent on food packaging and processing. The existing food packaging technologies are extremely popular because of their easy processing methods as well as low cost. New technologies should be tuned with the existing packaging and processing advantages to sustain in the market. Although synthetic plastics such as PP, HDPE, LDPE, PS, and PET are the best options to be used for food packaging due to their availability and inexpensive nature, a huge amount of money must be spent every year for their recycling programs, which is about $75-$209 per ton of plastic waste recycling and a total of approximately $624 billion in a year [143]. The recycling process includes the cost of collection, transportation of recycled plastics, processing, and treatments. Although, recycling processes are costly, still this is important to hinder tremendous ecological effects of these non-biodegradable plastics. Therefore, the goal of the new technologies is to define the standard and economically viable methods, set the guidelines of using plastic products, and find out new easy solutions of plastic recycling. The use of bioplastics in food packaging can be a way out to reduce the recycling cost which is advantageous both economically and environmentally.

CONCLUSION

This chapter offers a better understanding about the contributions of polymers and their nanocomposite materials in various fields of applications such as food packaging, electronic packaging, and industrial packaging. Polymer nanocomposites offer large variety of properties such as mechanical strength, gas barrier property, water vapour resistance, along with some special characteristics such as to prevent invasion of bacteria and microorganisms (antimicrobial effect), resist the formation of biofilm (antifouling property), *etc.* These properties are particularly important in case of preserving vegetables, fruits, beverages, bakery products, *etc.* Nowadays, along with the fossil-fuel derived polymers such as HDPE, LDPE, PS, PP, PET, PVC *etc.*, bio-derived polymers are also gaining immense importance in packaging industries due to their easy availability and biodegradability which impart less/ no harmful effects to the environment.

Due to the global environmental problems, currently, the utilization of the bio-based materials is the prime topic of interests in packaging research. However, industrial scale implementations of biopolymers are still limited due to the poor mechanical strength and gas barrier properties associated with these kinds of polymers. That is why frequently biopolymers are used together with the conventionally used fossil fuel-derived polymers to reduce the environmental toxicity of the packaging materials. Recently, attention has been paid to the fabrication of organic-inorganic hybrid composites based on bio-derived polymers. Among different nanofillers, layered silicates and graphene or graphene oxide are the most preferable for this purpose due to their good compatibility with polymers, excellent gas barrier property because of the formation of the tortuous path in the nanocomposite films, and bioactivity like antimicrobial property, which extends the shelf-life of the packaging material. Use of biopolymers in packaging would also reduce the waste generation.

Great attention has been paid to the fabrication of smart/ intelligent packaging materials so that any kind of changes in the food products such as pH, temperature, gas formation, biofilm formation, *etc.* can be detected. All these efforts will obviously pave a new era in the packaging industry in near future.

FUTURE SCOPE

The academic research and continuous improvement in the packaging technology have led to diverse advanced solutions to enhance the food-safety and the shelf-life of the packaged product. Although the use of active and smart packaging materials is very much exciting and they have tremendous prospects in near future, the same legislation of the conventional packaging should also be applicable for the active and smart packaging. However, till date, these kinds of intelligent packaging materials are in their very initial stage of utilization, and because of this, no standard method could have been developed to optimise the suitability of using active packaging in direct contact of foods. A major concern of this active packaging is the direct contact of the sensor, which is present in the packaging material with food. This will initiate the possible transmission of active ingredients to the food either intentionally or unintentionally along with subsequent probable health hazards. So, the standard should be implemented as soon as possible to the use of substances or the amount of substances in order to maintain health safety of the consumers. Additionally, the cost of packaging and availability of that product to all types of consumers should be synchronized so that a beneficial outcome can be obtained from the technological innovations. Although the attitude of industries and consumers is positive towards active and intelligent packaging, a mutual tunning should be developed between the manufacturers, traders, and consumers to successfully introduce this technology in a large scale. Fortunately, cheaper alternatives have been developed by 3D printing methods [144]. Finally, the technologists and manufacturers must assure the consumer about the safe use of these products as the existing market products.

ACKNOWLEDGEMENT

Declared none.

REFERENCES

[1]   Sharma C, Dhiman R, Rokana N, Panwar H. Nanotechnology: an untapped resource for food packaging. Front Microbiol 2017; 8: 1735.
[http://dx.doi.org/10.3389/fmicb.2017.01735] [PMID: 28955314]

[2]   Hatzigrigoriou NB, Papaspyrides CD. Nanotechnology in plastic food-contact materials. J Appl Polym Sci 2011; 122(6): 3719-38.
[http://dx.doi.org/10.1002/app.34786]

[3]   Rossi M, Passeri D, Sinibaldi A, *et al.* Nanotechnology for food packaging and food quality assessment. Adv Food Nutr Res 2017; 82: 149-204.
[http://dx.doi.org/10.1016/bs.afnr.2017.01.002] [PMID: 28427532]

[4]   Abbas K A, Saleh A M, Mohamed A. J Food Agric Environ 2009; 7(3–4): 14-7.

[5]   Wan YJ, Li G, Yao YM, Zeng XL, Zhu PL, Sun R. Recent advances in polymer-based electronic packaging materials. Compo Commun 2020; 19: 154-67.
[http://dx.doi.org/10.1016/j.coco.2020.03.011]

[6]   Singh A, Sharma PK, Malviya R. Eco friendly pharmaceutical packaging material. World Appl Sci J 2011; 14(11): 1703-16.

[7]   López-Rubio A, Almenar E, Hernandez-Muñoz P, Lagarón JM, Catalá R, Gavara R. Overview of active polymer-based packaging technologies for food applications. Food Rev Int 2004; 20(4): 357-87.
[http://dx.doi.org/10.1081/FRI-200033462]

[8]   Denault J, Labrecque B. Technology group on polymer nanocomposites–PNC-Tech. Canada: Ind. Mater. Institute. Natl. Res. Counc 2004; 75.

[9]   Kalendova A, Merinska D, Gerard JF, Slouf M. Polymer/clay nanocomposites and their gas barrier properties. Polym Compos 2013; 34(9): 1418-24.
        [http://dx.doi.org/10.1002/pc.22541]

[10]   Tan B, Thomas NL. A review of the water barrier properties of polymer/clay and polymer/graphene nanocomposites. J Membr Sci 2016; 514: 595-612.
        [http://dx.doi.org/10.1016/j.memsci.2016.05.026]

[11]   Youssef AM. Polymer nanocomposites as a new trend for packaging applications. Polym Plast Technol Eng 2013; 52(7): 635-60.
        [http://dx.doi.org/10.1080/03602559.2012.762673]

[12]   Siracusa V, Rosa MD. Sustainable packaging, in Sustainable Food Systems from Agriculture to Industry. Elsevier 2018; pp. 275-307.
        [http://dx.doi.org/10.1016/B978-0-12-811935-8.00008-1]

[13]   Scott G. Why degradable polymers? in Degradable Polymers. Springer 2002; pp. 1-15.
        [http://dx.doi.org/10.1007/978-94-017-1217-0]

[14]   Singh A, Banerjee SL, Kumari K, Kundu PP. Recent innovations in chemical recycling of polyethylene terephthalate waste: a circular economy approach towards sustainability, handb solid waste manag sustain through Circ. Econ 2020; pp. 1-28.

[15]   Malathi AN, Santhosh KS, Nidoni U. Recent trends of biodegradable polymer: biodegradable films for food packaging and application of nanotechnology in biodegradable food packaging. Curr Trends Technol Sci 2014; 3(2): 73-9.

[16]   Ozdemir M, Floros JD. Active food packaging technologies. Crit Rev Food Sci Nutr 2004; 44(3): 185-93.
        [http://dx.doi.org/10.1080/10408690490441578] [PMID: 15239372]

[17]   Li K, Jin S, Li J, Chen H. Improvement in antibacterial and functional properties of mussel-inspired cellulose nanofibrils/gelatin nanocomposites incorporated with graphene oxide for active packaging. Ind Crops Prod 2019; 132: 197-212.
        [http://dx.doi.org/10.1016/j.indcrop.2019.02.011]

[18]   Lau K, Gu C, Hui D. A critical review on nanotube and nanotube/nanoclay related polymer composite materials. Compos, Part B Eng 2006; 37(6): 425-36.
        [http://dx.doi.org/10.1016/j.compositesb.2006.02.020]

[19]   Singha S, Hedenqvist MS. A review on barrier properties of poly (lactic acid)/clay nanocomposites. Polymers (Basel) 2020; 12(5): 1095.
        [http://dx.doi.org/10.3390/polym12051095] [PMID: 32403371]

[20]   Behjat T, Arabkhedri M. Polymers and food packaging.Polymer science and innovative applications. Elsevier 2020; pp. 525-43.

[21]   Zhang M, Biesold GM, Choi W, *et al.* Recent advances in polymers and polymer composites for food packaging. Mater Today 2022; 53: 134-61.
        [http://dx.doi.org/10.1016/j.mattod.2022.01.022]

[22]   Girija BG, Sailaja RRN, Madras G. Thermal degradation and mechanical properties of PET blends. Polym Degrad Stabil 2005; 90(1): 147-53.
        [http://dx.doi.org/10.1016/j.polymdegradstab.2005.03.003]

[23]   Peltzer M, Wagner J, Jiménez A. Migration study of carvacrol as a natural antioxidant in high-density polyethylene for active packaging. Food Addit Contam Part A Chem Anal Control Expo Risk Assess 2009; 26(6): 938-46.
        [http://dx.doi.org/10.1080/02652030802712681] [PMID: 19680969]

[24] Pearson R. PVC as a food packaging material. Food Chem 1982; 8(2): 85-96.
[http://dx.doi.org/10.1016/0308-8146(82)90004-8]

[25] Azlin-Hasim S, Cruz-Romero MC, Cummins E, Kerry JP, Morris MA. The potential use of a layer-by-layer strategy to develop LDPE antimicrobial films coated with silver nanoparticles for packaging applications. J Colloid Interface Sci 2016; 461: 239-48.
[http://dx.doi.org/10.1016/j.jcis.2015.09.021] [PMID: 26402783]

[26] Lepot N, Van Bael MK, Van den Rul H, *et al.* Influence of incorporation of ZnO nanoparticles and biaxial orientation on mechanical and oxygen barrier properties of polypropylene films for food packaging applications. J Appl Polym Sci 2011; 120(3): 1616-23.
[http://dx.doi.org/10.1002/app.33277]

[27] Brighton C. Styrene polymers and food packaging. Food Chem 1982; 8(2): 97-107.
[http://dx.doi.org/10.1016/0308-8146(82)90005-X]

[28] Helmroth E, Rijk R, Dekker M, Jongen W. Predictive modelling of migration from packaging materials into food products for regulatory purposes. Trends Food Sci Technol 2002; 13(3): 102-9.
[http://dx.doi.org/10.1016/S0924-2244(02)00031-6]

[29] Kerry JP, O'Grady MN, Hogan SA. Past, current and potential utilisation of active and intelligent packaging systems for meat and muscle-based products: A review. Meat Sci 2006; 74(1): 113-30.
[http://dx.doi.org/10.1016/j.meatsci.2006.04.024] [PMID: 22062721]

[30] Hassan MES, Bai J, Dou D-Q. Biopolymers; definition, classification and applications. Egypt J Chem 2019; 62(9): 1725-37.

[31] Ishida H, Campbell S, Blackwell J. General approach to nanocomposite preparation. Chem Mater 2000; 12(5): 1260-7.
[http://dx.doi.org/10.1021/cm990479y]

[32] Huang YJ, Qin YW, Dong JY, Zhao X, Hu X. PE/OMMT nanocomposites prepared by *in situ* polymerization approach: Effects of OMMT-intercalated catalysts and silicate modifications. J Appl Polym Sci 2012; 123(5): 3106-16.
[http://dx.doi.org/10.1002/app.34939]

[33] López-Manchado MA, Herrero B, Arroyo M. Organoclay–natural rubber nanocomposites synthesized by mechanical and solution mixing methods. Polym Int 2004; 53(11): 1766-72.
[http://dx.doi.org/10.1002/pi.1573]

[34] Fox J. Analysis of polymer additives in the packaging industry. Florida Gainesville, FL, USA: Univ. 2008.

[35] Cherif Lahimer M, Ayed N, Horriche J, Belgaied S. Characterization of plastic packaging additives: Food contact, stability and toxicity. Arab J Chem 2017; 10: S1938-54.
[http://dx.doi.org/10.1016/j.arabjc.2013.07.022]

[36] Mekonnen T, Mussone P, Khalil H, Bressler D. Progress in bio-based plastics and plasticizing modifications. J Mater Chem A Mater Energy Sustain 2013; 1(43): 13379-98.
[http://dx.doi.org/10.1039/c3ta12555f]

[37] Kutz M. Applied plastics engineering handbook: processing and materials. William Andrew 2011.

[38] Singh A, Khamrai M, Samanta S, Kumari K, Kundu PP. Microbial, physicochemical, and sensory analyzes-based shelf life appraisal of white fresh cheese packaged into PET waste-based active packaging film. J Pack Techn Res 2018; 2(2): 125-47.
[http://dx.doi.org/10.1007/s41783-018-0034-5]

[39] Regulation C. No 10/2011 of 14 January 2011 on plastic materials and articles intended to come into contact with food 2011.

[40] Peters RJB, Bouwmeester H, Gottardo S, *et al.* Nanomaterials for products and application in agriculture, feed and food. Trends Food Sci Technol 2016; 54: 155-64.

[http://dx.doi.org/10.1016/j.tifs.2016.06.008]

[41]   Yemmireddy VK, Hung YC. Effect of binder on the physical stability and bactericidal property of titanium dioxide (TiO₂) nanocoatings on food contact surfaces. Food Control 2015; 57: 82-8.
[http://dx.doi.org/10.1016/j.foodcont.2015.04.009]

[42]   Amini E, Azadfallah M, Layeghi M, Talaei-Hassanloui R. Silver-nanoparticle-impregnated cellulose nanofiber coating for packaging paper. Cellulose 2016; 23(1): 557-70.
[http://dx.doi.org/10.1007/s10570-015-0846-1]

[43]   Annous BA, Fratamico PM, Smith JL. Scientific status summary. J Food Sci 2009; 74(1): R24-37.
[http://dx.doi.org/10.1111/j.1750-3841.2008.01022.x] [PMID: 19200115]

[44]   Ganguly S, Dana K, Mukhopadhyay TK, Parya TK, Ghatak S. Organophilic nano clay: A comprehensive review. Trans Indian Ceram Soc 2011; 70(4): 189-206.
[http://dx.doi.org/10.1080/0371750X.2011.10600169]

[45]   García-López D, Picazo O, Merino JC, Pastor JM. Polypropylene–clay nanocomposites: effect of compatibilizing agents on clay dispersion. Eur Polym J 2003; 39(5): 945-50.
[http://dx.doi.org/10.1016/S0014-3057(02)00333-6]

[46]   Farhoodi M. Nanocomposite materials for food packaging applications: characterization and safety evaluation. Food Eng Rev 2016; 8(1): 35-51.
[http://dx.doi.org/10.1007/s12393-015-9114-2]

[47]   Majeed K, Jawaid M, Hassan A, *et al.* Potential materials for food packaging from nanoclay/natural fibres filled hybrid composites. Mater Des 2013; 46: 391-410.
[http://dx.doi.org/10.1016/j.matdes.2012.10.044]

[48]   Kim SG, Lofgren EA, Jabarin SA. Dispersion of nanoclays with poly(ethylene terephthalate) by melt blending and solid state polymerization. J Appl Polym Sci 2013; 127(3): 2201-12.
[http://dx.doi.org/10.1002/app.37796]

[49]   Popov V, Hinkov I, Diankov S, Karsheva M, Handzhiyski Y. Ultrasound-assisted green synthesis of silver nanoparticles and their incorporation in antibacterial cellulose packaging. Green Processing and Synthesis 2015; 4(2): 125-31.
[http://dx.doi.org/10.1515/gps-2014-0085]

[50]   Yildirim S, Röcker B. Active packaging, in Nanomaterials for Food Packaging. Elsevier 2018; pp. 173-202.
[http://dx.doi.org/10.1016/B978-0-323-51271-8.00007-3]

[51]   O'Grady MN, Monahan FJ, Bailey J, Allen P, Buckley DJ, Keane MG. Colour-stabilising effect of muscle vitamin E in minced beef stored in high oxygen packs. Meat Sci 1998; 50(1): 73-80.
[http://dx.doi.org/10.1016/S0309-1740(98)00017-5] [PMID: 22060810]

[52]   Appendini P, Hotchkiss JH. Review of antimicrobial food packaging. Innov Food Sci Emerg Technol 2002; 3(2): 113-26.
[http://dx.doi.org/10.1016/S1466-8564(02)00012-7]

[53]   Knetsch MLW, Koole LH. New strategies in the development of antimicrobial coatings: the example of increasing usage of silver and silver nanoparticles. Polymers (Basel) 2011; 3(1): 340-66.
[http://dx.doi.org/10.3390/polym3010340]

[54]   Malhotra B, Keshwani A, Kharkwal H. Antimicrobial food packaging: potential and pitfalls. Front Microbiol 2015; 6: 611.
[http://dx.doi.org/10.3389/fmicb.2015.00611] [PMID: 26136740]

[55]   Fasihnia SH, Peighambardoust SH, Peighambardoust SJ. Nanocomposite films containing organoclay nanoparticles as an antimicrobial (active) packaging for potential food application. J Food Process Preserv 2018; 42(2): e13488.
[http://dx.doi.org/10.1111/jfpp.13488]

[56]   Yahiaoui F, Benhacine F, Ferfera-Harrar H, Habi A, Hadj-Hamou AS, Grohens Y. Development of antimicrobial PCL/nanoclay nanocomposite films with enhanced mechanical and water vapor barrier properties for packaging applications. Polym Bull 2015; 72(2): 235-54.
[http://dx.doi.org/10.1007/s00289-014-1269-0]

[57]   Hadj-Hamou AS, Metref F, Yahiaoui F. Thermal stability and decomposition kinetic studies of antimicrobial PCL/nanoclay packaging films. Polym Bull 2017; 74(9): 3833-53.
[http://dx.doi.org/10.1007/s00289-017-1929-y]

[58]   Liu G, Song Y, Wang J, *et al.* Effects of nanoclay type on the physical and antimicrobial properties of PVOH-based nanocomposite films. Lebensm Wiss Technol 2014; 57(2): 562-8.
[http://dx.doi.org/10.1016/j.lwt.2014.01.009]

[59]   Rhim JW, Hong SI, Ha CS. Tensile, water vapor barrier and antimicrobial properties of PLA/nanoclay composite films. Lebensm Wiss Technol 2009; 42(2): 612-7.
[http://dx.doi.org/10.1016/j.lwt.2008.02.015]

[60]   Shankar S, Chorachoo J, Jaiswal L, Voravuthikunchai SP. Effect of reducing agent concentrations and temperature on characteristics and antimicrobial activity of silver nanoparticles. Mater Lett 2014; 137: 160-3.
[http://dx.doi.org/10.1016/j.matlet.2014.08.100]

[61]   Rhim JW, Wang LF, Hong SI. Preparation and characterization of agar/silver nanoparticles composite films with antimicrobial activity. Food Hydrocoll 2013; 33(2): 327-35.
[http://dx.doi.org/10.1016/j.foodhyd.2013.04.002]

[62]   Shankar S, Teng X, Rhim JW. Properties and characterization of agar/CuNP bionanocomposite films prepared with different copper salts and reducing agents. Carbohydr Polym 2014; 114: 484-92.
[http://dx.doi.org/10.1016/j.carbpol.2014.08.036] [PMID: 25263917]

[63]   Kanmani P, Rhim JW. Physical, mechanical and antimicrobial properties of gelatin based active nanocomposite films containing AgNPs and nanoclay. Food Hydrocoll 2014; 35: 644-52.
[http://dx.doi.org/10.1016/j.foodhyd.2013.08.011]

[64]   Azam A, Ahmed AS, Oves M, Khan MS, Habib SS, Memic A. Antimicrobial activity of metal oxide nanoparticles against Gram-positive and Gram-negative bacteria: a comparative study. Int J Nanomedicine 2012; 7: 6003-9.
[http://dx.doi.org/10.2147/IJN.S35347] [PMID: 23233805]

[65]   Lagaron JM, Catalá R, Gavara R. Structural characteristics defining high barrier properties in polymeric materials. Mater Sci Technol 2004; 20(1): 1-7.
[http://dx.doi.org/10.1179/026708304225010442]

[66]   Teixeira V, Carneiro J, Carvalho P, Silva E, Azevedo S, Batista C. High barrier plastics using nanoscale inorganic films, in Multifunctional and nanoreinforced polymers for food packaging. Elsevier 2011; pp. 285-315.
[http://dx.doi.org/10.1533/9780857092786.1.285]

[67]   Imran M, Revol-Junelles AM, Martyn A, *et al.* Active food packaging evolution: transformation from micro- to nanotechnology. Crit Rev Food Sci Nutr 2010; 50(9): 799-821.
[http://dx.doi.org/10.1080/10408398.2010.503694] [PMID: 20924864]

[68]   Janjarasskul T, Tananuwong K, Krochta JM. Whey protein film with oxygen scavenging function by incorporation of ascorbic acid. J Food Sci 2011; 76(9): E561-8.
[http://dx.doi.org/10.1111/j.1750-3841.2011.02409.x] [PMID: 22416701]

[69]   Xiao-e L, Green ANM, Haque SA, Mills A, Durrant JR. Light-driven oxygen scavenging by titania/polymer nanocomposite films. J Photochem Photobiol Chem 2004; 162(2-3): 253-9.
[http://dx.doi.org/10.1016/j.nainr.2003.08.010]

[70]   Maneerat C, Hayata Y. Gas-phase photocatalytic oxidation of ethylene with TiO_2-coated packaging film for horticultural products. Trans ASABE 2008; 51(1): 163-8.

[http://dx.doi.org/10.13031/2013.24200]

[71]   Majid I, Ahmad Nayik G, Mohammad Dar S, Nanda V. Novel food packaging technologies: Innovations and future prospective. J Saudi Soc Agric Sci 2018; 17(4): 454-62.
[http://dx.doi.org/10.1016/j.jssas.2016.11.003]

[72]   Ashfaq A, Khursheed N, Fatima S, Anjum Z, Younis K. Application of nanotechnology in food packaging: Pros and Cons. Journal of Agriculture and Food Research 2022; 7: 100270.
[http://dx.doi.org/10.1016/j.jafr.2022.100270]

[73]   Aliofkhazraei M. Synthesis, processing and application of nanostructured coatings, in Nanocoatings. Springer 2011; pp. 1-28.

[74]   Duncan TV. Applications of nanotechnology in food packaging and food safety: Barrier materials, antimicrobials and sensors. J Colloid Interface Sci 2011; 363(1): 1-24.
[http://dx.doi.org/10.1016/j.jcis.2011.07.017] [PMID: 21824625]

[75]   Selke S. Nanotechnology and packaging.Wiley Encycl Packag Technol. 3rd ed. Hoboken, NJ: John Wiley Sons, Inc 2009; pp. 813-8.

[76]   Applerot G, Perkas N, Amirian G, Girshevitz O, Gedanken A. Coating of glass with ZnO *via* ultrasonic irradiation and a study of its antibacterial properties. Appl Surf Sci 2009; 256(3): S3-8.
[http://dx.doi.org/10.1016/j.apsusc.2009.04.198]

[77]   Chawengkijwanich C, Hayata Y. Development of TiO_2 powder-coated food packaging film and its ability to inactivate *Escherichia coli in vitro* and in actual tests. Int J Food Microbiol 2008; 123(3): 288-92.
[http://dx.doi.org/10.1016/j.ijfoodmicro.2007.12.017] [PMID: 18262298]

[78]   Vartiainen J, Tuominen M, Nättinen K. Bio-hybrid nanocomposite coatings from sonicated chitosan and nanoclay. J Appl Polym Sci 2010; 116(6): NA.
[http://dx.doi.org/10.1002/app.31922]

[79]   Travan A, Marsich E, Donati I, *et al.* Silver–polysaccharide nanocomposite antimicrobial coatings for methacrylic thermosets. Acta Biomater 2011; 7(1): 337-46.
[http://dx.doi.org/10.1016/j.actbio.2010.07.024] [PMID: 20656078]

[80]   Shemesh R, Krepker M, Goldman D, *et al.* Antibacterial and antifungal LDPE films for active packaging. Polym Adv Technol 2015; 26(1): 110-6.
[http://dx.doi.org/10.1002/pat.3434]

[81]   Lvov YM, Shchukin DG, Möhwald H, Price RR. Halloysite clay nanotubes for controlled release of protective agents. ACS Nano 2008; 2(5): 814-20.
[http://dx.doi.org/10.1021/nn800259q] [PMID: 19206476]

[82]   Biddeci G, Cavallaro G, Di Blasi F, *et al.* Halloysite nanotubes loaded with peppermint essential oil as filler for functional biopolymer film. Carbohydr Polym 2016; 152: 548-57.
[http://dx.doi.org/10.1016/j.carbpol.2016.07.041] [PMID: 27516303]

[83]   Tornuk F, Sagdic O, Hancer M, Yetim H. Development of LLDPE based active nanocomposite films with nanoclays impregnated with volatile compounds. Food Res Int 2018; 107: 337-45.
[http://dx.doi.org/10.1016/j.foodres.2018.02.036] [PMID: 29580493]

[84]   Khan B, Bilal Khan Niazi M, Samin G, Jahan Z. Thermoplastic starch: a possible biodegradable food packaging material—a review. J Food Process Eng 2017; 40(3): e12447.
[http://dx.doi.org/10.1111/jfpe.12447]

[85]   Siracusa V, Rocculi P, Romani S, Rosa MD. Biodegradable polymers for food packaging: a review. Trends Food Sci Technol 2008; 19(12): 634-43.
[http://dx.doi.org/10.1016/j.tifs.2008.07.003]

[86]   Shatalov I, Shatalova A, Shleikin A. Developing of edible packaging material based on protein film

[87]   Gheorghita Puscaselu R, Amariei S, Norocel L, Gutt G. New edible packaging material with function

in shelf life extension: applications for the meat and cheese industries. Foods 2020; 9(5): 562.
[http://dx.doi.org/10.3390/foods9050562] [PMID: 32370262]

[88]   K. Huff, Active and intelligent packaging: innovations for the future, Dep. Food Sci. Technol. Virginia Polytech. Inst. State Univ. Blacksburg, Va, 2008, pp. 1–13.

[89]   Bumbudsanpharoke N, Ko S. Nanomaterial-based optical indicators: Promise, opportunities, and challenges in the development of colorimetric systems for intelligent packaging. Nano Res 2019; 12(3): 489-500.
[http://dx.doi.org/10.1007/s12274-018-2237-z]

[90]   Alberti G, Zanoni C, Magnaghi LR, Biesuz R. Gold and silver nanoparticle-based colorimetric sensors: new trends and applications. Chemosensors (Basel) 2021; 9(11): 305.
[http://dx.doi.org/10.3390/chemosensors9110305]

[91]   J. H. Han, C. H. L. Ho, and E. T. Rodrigues, Intelligent packaging, Innov. food Packag. 2005; 138-155.
[http://dx.doi.org/10.1016/B978-012311632-1/50041-3]

[92]   Sanvicens N, Pastells C, Pascual N, Marco MP. Nanoparticle-based biosensors for detection of pathogenic bacteria. Trends Analyt Chem 2009; 28(11): 1243-52.
[http://dx.doi.org/10.1016/j.trac.2009.08.002]

[93]   Pérez-López B, Merkoçi A. Nanomaterials based biosensors for food analysis applications. Trends Food Sci Technol 2011; 22(11): 625-39.
[http://dx.doi.org/10.1016/j.tifs.2011.04.001]

[94]   De Jong AR, Boumans H, Slaghek T, Van Veen J, Rijk R, Van Zandvoort M. Active and intelligent packaging for food: Is it the future? Food Addit Contam 2005; 22(10): 975-9.
[http://dx.doi.org/10.1080/02652030500336254] [PMID: 16227181]

[95]   Pavase TR, Lin H, Shaikh Q, *et al.* Recent advances of conjugated polymer (CP) nanocomposite-based chemical sensors and their applications in food spoilage detection: A comprehensive review. Sens Actuators B Chem 2018; 273: 1113-38.
[http://dx.doi.org/10.1016/j.snb.2018.06.118]

[96]   Vanderroost M, Ragaert P, Devlieghere F, De Meulenaer B. Intelligent food packaging: The next generation. Trends Food Sci Technol 2014; 39(1): 47-62.
[http://dx.doi.org/10.1016/j.tifs.2014.06.009]

[97]   Sadoh A, Hossain S, Ravindra NM. Thermochromic polymeric films for applications in active intelligent packaging—an overview. Micromachines (Basel) 2021; 12(10): 1193.
[http://dx.doi.org/10.3390/mi12101193] [PMID: 34683245]

[98]   Bäckman M. Feasibility study of thermochromic inks for the packaging industry 2017.

[99]   Pavelková A. Time temperature indicators as devices intelligent packaging. Acta Univ Agric Silvic Mendel Brun 2013; 61(1): 245-51.
[http://dx.doi.org/10.11118/actaun201361010245]

[100] Dobrucka R, Cierpiszewski R. Active and intelligent packaging food-Research and development-A Review. Pol J Food Nutr Sci 2014; 64(1): 7-15.
[http://dx.doi.org/10.2478/v10222-012-0091-3]

[101] Wang S, Liu X, Yang M, Zhang Y, Xiang K, Tang R. Review of time temperature indicators as quality monitors in food packaging. Packag Technol Sci 2015; 28(10): 839-67.
[http://dx.doi.org/10.1002/pts.2148]

[102] Endoza TFM, Welt BA, Otwell S, Teixeira AA, Kristonsson H, Balaban MO. Kinetic parameter estimation of time-temperature integrators intended for use with packaged fresh seafood. J Food Sci 2004; 69(3): FMS90-6.
[http://dx.doi.org/10.1111/j.1365-2621.2004.tb13377.x]

[103] Kuswandi B. Freshness sensors for food packaging. Ref Modul food Sci 2017; 1-11.
[http://dx.doi.org/10.1016/B978-0-08-100596-5.21876-3]

[104] Shao P, Liu L, Yu J, *et al.* An overview of intelligent freshness indicator packaging for food quality and safety monitoring. Trends Food Sci Technol 2021; 118: 285-96.
[http://dx.doi.org/10.1016/j.tifs.2021.10.012]

[105] Gutiérrez TJ, Ponce AG, Alvarez VA. Nano-clays from natural and modified montmorillonite with and without added blueberry extract for active and intelligent food nanopackaging materials. Mater Chem Phys 2017; 194: 283-92.
[http://dx.doi.org/10.1016/j.matchemphys.2017.03.052]

[106] Pirsa S, Karimi Sani I, Khodayvandi S. Design and fabrication of starch-nano clay composite films loaded with methyl orange and bromocresol green for determination of spoilage in milk package. Polym Adv Technol 2018; 29(11): 2750-8.
[http://dx.doi.org/10.1002/pat.4397]

[107] Gander P. The smart money is on intelligent design. UK: Food Manuf 2007; 1.

[108] Ahmed I, Lin H, Zou L, *et al.* An overview of smart packaging technologies for monitoring safety and quality of meat and meat products. Packag Technol Sci 2018; 31(7): 449-71.
[http://dx.doi.org/10.1002/pts.2380]

[109] Yam KL. Intelligent packaging for the future smart kitchen, Packag. Technol. Sci. Int J 2000; 13(2): 83-5.

[110] Nelson KA, Labuza TP. An evaluation of the kinetics of microwave doneness indicators. J Food Prot 1992; 55(3): 203-7.
[http://dx.doi.org/10.4315/0362-028X-55.3.203] [PMID: 31071836]

[111] Yam KL, Takhistov PT, Miltz J. Intelligent packaging: concepts and applications. J Food Sci 2005; 70(1): R1-R10.
[http://dx.doi.org/10.1111/j.1365-2621.2005.tb09052.x]

[112] Skura BJ. Modified atmosphere packaging of fish and fish products. Modif. Atmos. Packag. Food 1991; pp. 148-68.
[http://dx.doi.org/10.1007/978-1-4615-2117-4_6]

[113] Smith JP, Hoshino J, Abe Y. Interactive packaging involving sachet technology, in Active food packaging. Springer 1995; pp. 143-73.
[http://dx.doi.org/10.1007/978-1-4615-2175-4_6]

[114] Gorris LGM, Peppelenbos HW. Modified atmosphere and vacuum packaging to extend the shelf life of respiring food products. Horttechnology 1992; 2(3): 303-9.
[http://dx.doi.org/10.21273/HORTTECH.2.3.303]

[115] Sobhan A, Muthukumarappan K, Wei L, Van Den Top T, Zhou R. Development of an activated carbon-based nanocomposite film with antibacterial property for smart food packaging. Mater Today Commun 2020; 23: 101124.
[http://dx.doi.org/10.1016/j.mtcomm.2020.101124]

[116] Bratovčić A, Odobašić A, Ćatić S, Šestan I. Application of polymer nanocomposite materials in food packaging. Croat J Food Sci Technol 2015; 7(2): 86-94.
[http://dx.doi.org/10.17508/CJFST.2015.7.2.06]

[117] Page K, Palgrave RG, Parkin IP, Wilson M, Savin SLP, Chadwick AV. Titania and silver–titania composite films on glass—potent antimicrobial coatings. J Mater Chem 2007; 17(1): 95-104.
[http://dx.doi.org/10.1039/B611740F]

[118] Oms-Oliu G, Soliva-Fortuny R, Martín-Belloso O. Physiological and microbiological changes in fresh-cut pears stored in high oxygen active packages compared with low oxygen active and passive modified atmosphere packaging. Postharvest Biol Technol 2008; 48(2): 295-301.

[http://dx.doi.org/10.1016/j.postharvbio.2007.10.002]

[119]   Youssef AM, Ibrahim MM, Selim AE, Bayoumy TM, Fehead TM. The influence of some antimicrobial nanoparticles blends on shelf-life of minced meat. Trans Egypt Soc Chem Eng 2011; 37: 36-55. [TESCE].

[120]   Hong SI, Rhim JW. Antimicrobial activity of organically modified nano-clays. J Nanosci Nanotechnol 2008; 8(11): 5818-24.
[http://dx.doi.org/10.1166/jnn.2008.248] [PMID: 19198311]

[121]   Cha DS, Chinnan MS. Biopolymer-based antimicrobial packaging: a review. Crit Rev Food Sci Nutr 2004; 44(4): 223-37.
[http://dx.doi.org/10.1080/10408690490464276] [PMID: 15462127]

[122]   Emamifar A, Kadivar M, Shahedi M, Soleimanian-Zad S. Effect of nanocomposite packaging containing Ag and ZnO on inactivation of *Lactobacillus plantarum* in orange juice. Food Control 2011; 22(3-4): 408-13.
[http://dx.doi.org/10.1016/j.foodcont.2010.09.011]

[123]   Souza MCC, Benassi MT, Meneghel RFA, Silva RSSF. Stability of unpasteurized and refrigerated orange juice. Braz Arch Biol Technol 2004; 47(3): 391-7.
[http://dx.doi.org/10.1590/S1516-89132004000300009]

[124]   Friend RH, Gymer RW, Holmes AB, *et al.* Electroluminescence in conjugated polymers. Nature 1999; 397(6715): 121-8.
[http://dx.doi.org/10.1038/16393]

[125]   Lappas A, Zorko A, Wortham E, *et al.* Low-energy magnetic excitations and morphology in layered hybrid perovskite−poly(dimethylsiloxane) nanocomposites. Chem Mater 2005; 17(5): 1199-207.
[http://dx.doi.org/10.1021/cm048744p]

[126]   Haroun AA, Youssef AM. Synthesis and electrical conductivity evaluation of novel hybrid poly (methyl methacrylate)/titanium dioxide nanowires. Synth Met 2011; 161(19-20): 2063-9.
[http://dx.doi.org/10.1016/j.synthmet.2011.07.011]

[127]   Youssef AM, El-Samahy MA, Abdel Rehim MH. Preparation of conductive paper composites based on natural cellulosic fibers for packaging applications. Carbohydr Polym 2012; 89(4): 1027-32.
[http://dx.doi.org/10.1016/j.carbpol.2012.03.044] [PMID: 24750909]

[128]   Nassar MA, Youssef AM. Mechanical and antibacterial properties of recycled carton paper coated by PS/Ag nanocomposites for packaging. Carbohydr Polym 2012; 89(1): 269-74.
[http://dx.doi.org/10.1016/j.carbpol.2012.03.007] [PMID: 24750633]

[129]   Liu Y, Chakrabartty S, Alocilja EC. Fundamental building blocks for molecular biowire based forward error-correcting biosensors. Nanotechnology 2007; 18(42): 424017.
[http://dx.doi.org/10.1088/0957-4484/18/42/424017] [PMID: 21730450]

[130]   Hamid SH. Handbook of polymer degradation. CRC Press 2000.
[http://dx.doi.org/10.1201/9781482270181]

[131]   Murphy J. Other types of additive: miscellaneous additives. Addit. Plast. Handb 2001; pp. 219-29.

[132]   Birley A. Plastics used in food packaging and the rôle of additives. Food Chem 1982; 8(2): 81-4.
[http://dx.doi.org/10.1016/0308-8146(82)90003-6]

[133]   Ahmad M, Bajahlan AS. Leaching of styrene and other aromatic compounds in drinking water from PS bottles. J Environ Sci (China) 2007; 19(4): 421-6.
[http://dx.doi.org/10.1016/S1001-0742(07)60070-9] [PMID: 17915704]

[134]   Schmid P, Kohler M, Meierhofer R, Luzi S, Wegelin M. Does the reuse of PET bottles during solar water disinfection pose a health risk due to the migration of plasticisers and other chemicals into the water? Water Res 2008; 42(20): 5054-60.
[http://dx.doi.org/10.1016/j.watres.2008.09.025] [PMID: 18929387]

[135] Rosca ID, Vergnaud JM. Approach for a testing system to evaluate food safety with polymer packages. Polym Test 2006; 25(4): 532-43.
[http://dx.doi.org/10.1016/j.polymertesting.2006.02.006]

[136] Begley T, Castle L, Feigenbaum A, *et al.* Evaluation of migration models that might be used in support of regulations for food-contact plastics. Food Addit Contam 2005; 22(1): 73-90.
[http://dx.doi.org/10.1080/02652030400028035] [PMID: 15895614]

[137] Pennarun PY, Dole P, Feigenbaum A. Functional barriers in PET recycled bottles. Part I. Determination of diffusion coefficients in bioriented PET with and without contact with food simulants. J Appl Polym Sci 2004; 92(5): 2845-58.
[http://dx.doi.org/10.1002/app.20202]

[138] Fletcher BL, Mackay ME. A model of plastics recycling: Does recycling reduce the amount of waste? Resour Conserv Recycling 1996; 17(2): 141-51.
[http://dx.doi.org/10.1016/0921-3449(96)01068-3]

[139] Sarker M, Rashid M M, Molla M. Waste plastic conversion into chemical product like naphtha 2011.
[http://dx.doi.org/10.4303/jfrea/R110101]

[140] Singh RK, Datta M, Nema AK. A new system for groundwater contamination hazard rating of landfills. J Environ Manage 2009; 91(2): 344-57.
[http://dx.doi.org/10.1016/j.jenvman.2009.09.003] [PMID: 19836127]

[141] Sidique SF, Lupi F, Joshi SV. The effects of behavior and attitudes on drop-off recycling activities. Resour Conserv Recycling 2010; 54(3): 163-70.
[http://dx.doi.org/10.1016/j.resconrec.2009.07.012]

[142] Evison T, Read AD. Local Authority recycling and waste — awareness publicity/promotion. Resour Conserv Recycling 2001; 32(3-4): 275-91.
[http://dx.doi.org/10.1016/S0921-3449(01)00066-0]

[143] Kinnaman TC. The costs of municipal curbside recycling and waste collection. Resour. Conserv. Recycl 2010; p. 864.

[144] Kimionis J, Isakov M, Koh BS, Georgiadis A, Tentzeris MM. 3D-printed origami packaging with inkjet-printed antennas for RF harvesting sensors. IEEE Trans Microw Theory Tech 2015; 63(12): 4521-32.
[http://dx.doi.org/10.1109/TMTT.2015.2494580]

CHAPTER 3

Polymer Composites in Tissue Engineering

Togam Ringu[1], **Sampad Ghosh**[2] and **Nabakumar Pramanik**[1,*]

[1] *Department of Chemistry, National Institute of Technology, Arunachal Pradesh, Arunachal Pradesh-791113, India*

[2] *Department of Chemistry, Nalanda College of Engineering, Nalanda-803108, Bihar, India*

Abstract: A composite is a multiphase material made of layers of stacked phase *i.e.,* a matrix, an interface and a reinforced phase. The matrix phase is the main constituent of a composite. The interface binds the matrix and the reinforced phase, whereas, the latter provides strength to the material. Based on the matrix and the reinforced phase, it may be classified into various types such as fibers, particles, polymers, ceramics and metals. Polymer composite is a sub-type of composite having a polymer matrix and different reinforced materials. Due to its biocompatible nature, it is widely used in the field of biomedical applications. Many manufacturing methods are used in composites, but some of the commonly used manufacturing techniques include hand lay-up, reinforced reaction injection molding (RRIM), centrifugal casting, *etc*. High strength, and ductility with lightweight, cytocompatibility, and non-toxicity are some of the properties due to which composite materials are widely used in various industries such as automobile, aerospace, sports equipment, and tissue engineering. In tissue engineering (TE), a biomaterial called a scaffold, is developed that evolves into a functional tissue. Enhanced cell proliferation, cell adhesion and cell viability are observed with the composite-developed scaffold. Scaffold is fabricated using two types of composites; natural and synthetic composites. The applications of polymer composites at the bioengineering level are of great interest nowadays. This chapter intends to study various physicochemical properties of polymer composites including their bioengineering/tissue engineering applications elaborately. The study investigating the physicochemical properties and bioengineering/tissue engineering applications of polymer composites may bestow valuable insight into the potential of polymer composites in modern science.

Keywords: Cytocompatibility, Cell proliferation, Interface, Matrix, Multiphase, Polymer composites, Reinforcements, Scaffolds, Tissue engineering.

INTRODUCTION

In the field of material science, polymer composite materials have shown great development and growth over the past decades due to their diverse characteristics

---

[*] **Corresponding author Nabakumar Pramanik:** Department of Chemistry, National Institute of Technology, Arunachal Pradesh, Arunachal Pradesh-791113, India; E-mail: pramaniknaba@gmail.com

Subhendu Bhandari, Prashant Gupta and Ayan Dey (Eds.)

and applications. From being used in aerospace and electronics industries to biomedical applications, these are some of the transitions seen over the years in the development of polymer composites [1]. The dynamic transition in the usage of polymer composite is shown in Fig. (**1**). Polymer composite materials have replaced the conventional use of materials *e.g.,* metals in automobiles are being replaced by carbon fiber composites which are lighter in weight and have higher strength, and improved speed and fuel efficiency when compared to conventionally used materials in automobiles [2]. Polymers are easy to manufacture and process and are very cost-effective compared to the conventionally used other materials. A polymer composite is defined as a combination of a polymer with non-polymeric components such as metals, particles, ceramic, fiber, *etc.* All these components when mixed together, form a polymer composite [3]. Each and every attribute used in polymeric composites gives equal potential to improve the quality and properties of polymeric composites. The constituents of polymer composites are of three phases [3]:

Matrix Phase

This is the main constituent of a polymer composite. It is continuous in nature and surrounds the other phase *i.e.,* the dispersed phase. The materials used in the matrix phase are usually ductile or tough in nature. It also transfers stress to the fiber and maintains stability. It protects reinforcement from environmental factors such as chemicals and moisture, and also the surfaces of the fibers from mechanical degradation. Examples of the matrix phases are polymer matrix, ceramic matrix, metal matrix, *etc.*

Reinforced Phase

A material that gives strength to the composite is known as a reinforced material. Reinforced materials are usually strong with low density. It can also provide thermal and electrical conductivity apart from structural properties. Examples of reinforced phases are fiber, particles, graphite, *etc.*

Interface

Interface is a layer that separates matrix and reinforcement, and binds or holds the two phases together through bonding or adhesion.

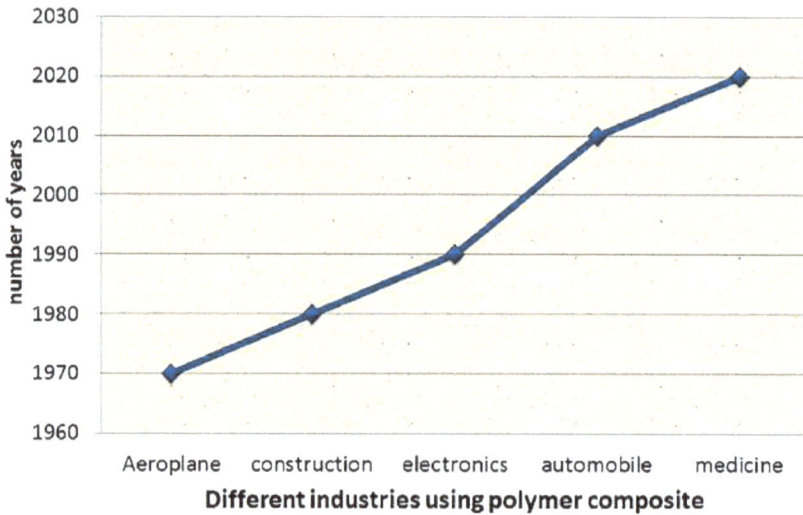

Fig. (1). Graphical representation of usage of polymer composite over the years.

Mixing or a combination of matrix and reinforced material is used in various *in-situ* and *ex-situ* methods. Mixing of these two phases creates a bond between them such as a mechanical bond or a chemical bond in which the matrix and the reinforced material are bound naturally or by using coupling agents to enhance their interactions with each other. Some of the commonly known coupling agents are silanes and titanates. In mechanical bonding, the roughness of the surface of fiber causes interlocking between the fiber and the matrix which causes mechanical bonding between them. Another method of bonding is reaction bonding; it happens when molecules of the fiber and matrix diffuse into each other at the interface. Thus, an interfacial layer is formed which is known as an interphase, and this helps in bonding between the two phases. Many new fabrications or modifications have been made in the concept of polymer composite in which it is modified or functionalized with certain components, which enhance the properties of the composite and reduce the cost of the product. Likewise, nowadays instead of conventional fibers, natural fibers are used as reinforcing materials which reduce the cost, improve the high specific properties, and have low density compared to conventional fiber materials [4]. Also, the usage of natural fibers leads to the manufacturing of biodegradable and non-abrasive attributes, unlike other reinforced fibers. Some of the common physicochemical properties of the composite materials are discussed in Table **1** referred to as [3].

Table 1. Characteristics of composites.

Sl. No.	Characteristics of Composites
1	Strength (stress which other materials cannot endure).
2	Stiffness (resistance of material from structural deformation).
3	Wear and corrosive resistance.
4	Chemical and fire resistance.
5	Abrasion resistance and weight reduction.
6	Thermal conductivity and acoustical insulation.
7	Fatigue resistance.

Functionalized polymer nanocomposites are also in trend nowadays, which show very good applications in the field of biomedicine, tissue engineering, drug delivery, bioimaging, *etc*. In a study, polymer nanocomposites were functionalized or combined with a metal-organic framework (MOF) which increased colloidal stability without the loss of crystallinity, increased targeting functionalities, enhanced stability and improved the application in drug delivery and bio-imaging devices [5].

In this book chapter, we are going to discuss the facts about the characteristics, classifications and types of various synthesis methods and the applications of polymer composites in the field of tissue engineering. Different combinations of polymer composites with biological and synthetic materials are generally used in tissue engineering.

CLASSIFICATION OF COMPOSITE

A polymer composite is classified into two types: reinforced materials and matrix materials. They are mentioned below:

Reinforced Materials

Reinforced materials give mechanical support to the composite and provide strength and stiffness to the composite which make them a very important component of a composite. The commonly used reinforced materials are natural fiber, carbon fiber, glass fiber, silica, clay, *etc*. The reinforced material composites may be subdivided into three types on the basis of different reinforced materials used which are schematically represented in Fig. (**2**).

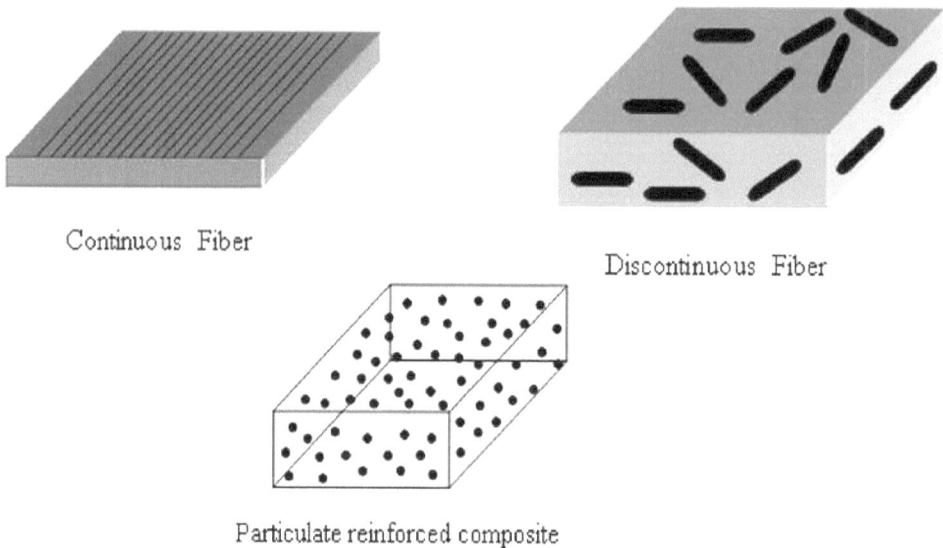

Fig. (2). Fiber and particulate reinforced composites.

Fiber Reinforced Composites

In fiber-reinforced composite (FRC), fiber is used as the reinforced material which gives mechanical strength to the composite. The most extensively used composite is the fiber-reinforced composite due to its sustainable properties [6]. When FRC is synthesized upon heating above the temperature of 200 ^{0}C, it becomes unstable. However, it can be rectified by several methods such as surface modification, chemical modification, plasma treatment, *etc.* Several different manufacturing techniques are used such as hand-layup, compression molding, resin transfer molding, automated fiber placement, and extrusion method [7]. The mechanical properties of the fiber-reinforced composites depend on the fiber strength, matrix strength, and the potential of interfacial adhesion between the fiber and the matrix [3]. The properties of FRC are high modulus and strength, thermal resistance, light-weight, abrasion resistance, corrosive resistance, fatigue resistance, excellent weathering stabilities, chemical resistance, UV resistance and electrical insulation [6]. Due to all these properties, it has wide applications such as in the aerospace industry, military sector, sports equipment, automotive industry, and also the shipment industry [8]. Common examples of the fibers in FRC are glass fiber, carbon fiber, Kevlar fiber, ultra-high molecular weight polyethylene fiber (UHMWPE fibers), poly (*p*-phenylene benzobisoxazole) fibers and basalt fibers [9]. A fiber-reinforced composite can be classified into two types based on the orientation of the fiber, *i.e.,* continuous and discontinuous fiber. A comparison between these two types is explained in Table **2** referred to as [10].

Table 2. Comparison between continuous and discontinuous fibers.

Sl. No.	Continuous Fiber	Discontinuous Fiber
1	It has a long aspect ratio *i.e.,* length to diameter ratio.	It has a short aspect ratio.
2	Aligned orientation.	Random orientation.
3	Strength and stiffness are higher than discontinuous fiber material.	Strength and stiffness are lower than the continuous fiber material.
4	Manufacturing cost is high.	Manufacturing cost is low.
5	Example: uni-directional, woven cloth and helical winding.	Example: chopped fiber and random mat.

Particle Reinforced Composites

Particle-reinforced composites use micro/nano-sized particles as the reinforced material. The shape, size, configuration and geometry of the particles are the key factors in determining and amplifying the mechanical properties of the composite. Particles may generally have a spherical, polyhedral, ellipsoidal and irregular shape. This type of composite is prepared by dispersing the reinforcing particles into matrices. This kind of composite is fabricated when the requirement is high wear resistance at a low cost. Particle-reinforced composites have less strength than fiber-reinforced composites. The distribution and dispersion of the particles play a crucial role in shaping the mechanical properties of the composites [11]. Hence, it is very important to choose a good and effective synthesis method. Good synthesis and fabrication methods provide enough shear force to overcome the adhesion of particle agglomerates and break it down into fine minor particle components as shown in Fig. (**2**) and Fig. (**3**).

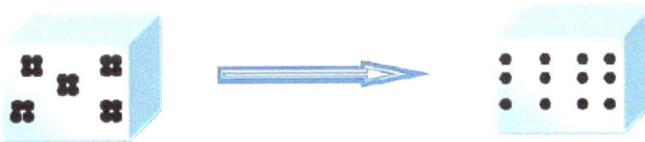

Fig. (3). Break down of particle agglomeration to a uniform distribution.

The particles used in the formation of a composite have to be broken down from the aggregates into minor components because uniformly distributed particles on the surface of the composite will give better mechanical properties and stability. In contrast to that, if the aggregates of a particle are not distributed or scattered properly, then the particles will form a cluster on the composite. Upon external force or stress, it will crack and result in premature failure of the composite [11]. Particle-reinforced composites can be subdivided into two types [12]: flake and filled/skeletal (Fig. **4**).

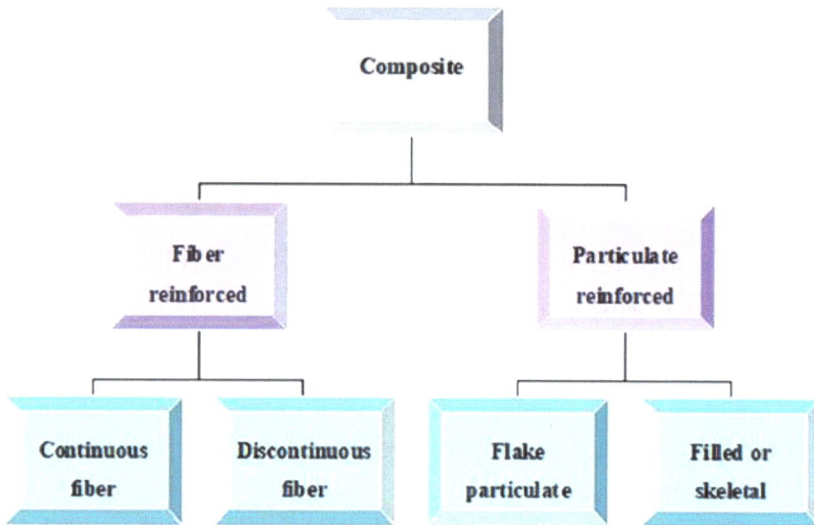

Fig. (4). Classification of composite based on reinforced material.

On the Basis of Matrix Materials

In a composite, the matrix should always be more than 50% of the material used as it helps in maintaining the stability of the composite. Based on the matrix material used in the composite, it can be classified into three types, *i.e.,* polymer matrix composite, metal matrix composite, and ceramic matrix composite. The polymer matrix is further subdivided into two more subclasses, *i.e.,* thermosets and thermoplastic. The classification is discussed in more detail below and the classification of matrix material is explained in Fig. (**5**).

Polymer Matrix Composites (PMC)

The usage of the polymer as a matrix material in a composite is known as polymer matrix composite. The primary phase is the polymer and the secondary phase is either fiber or particulate. PMC is more widely used compared to MMC and CMC because it is cost-effective and has an easy synthesis method. Moreover, the usage of a polymer as a matrix material has several upper hands over the other materials due to its low cost, easy processing steps, good chemical resistance, and low specific gravity [13]. The advantages and disadvantages of PMC are discussed in Table **3**.

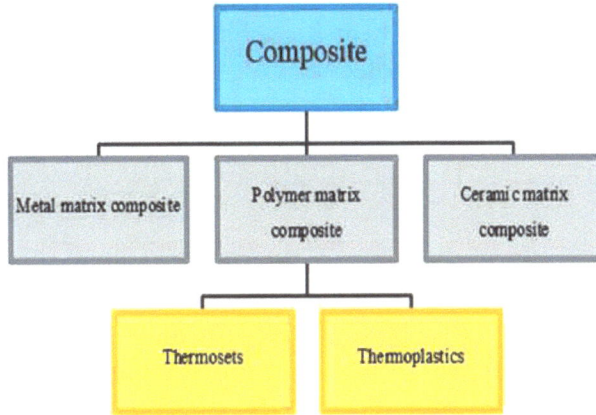

Fig. (5). Classification of composite based on the matrix.

Table 3. Advantages and disadvantages of polymer matrix composites.

Advantages	Disadvantages
• High sturdiness. • Highly rigid. • Fracture resistance. • Abrasion resistance. • Impact resistance. • Corrosion resistance. • Fatigue resistance. • Low cost.	• Low heat resistance. • High coefficient of thermal expansion.

There are several parameters in determining the characteristics of PMCs, and they are:

• Interfacial bonding between the polymer matrix and the reinforcement material.
• Geometry, orientation, and dispersed phase of the reinforced material [14].
• Properties of the polymer used in the matrix, *i.e.,* whether it is synthetic or natural [15].

The most commonly used polymer class is resin; apart from that polyester, epoxy, polyamide, polyethylene (PE) *etc.* are also used. And, also in recent years, the usage of natural polymers has increased because of their biocompatibility attributes, and some examples are collagen, gelatin, chitosan, alginate, hyaluronic acid, *etc.* PMC can be classified based on the matrix, *i.e.,* thermoset and thermoplastic [16]. A thermoset polymer is a polymer that is derived from the exchange of electrons by chemical bonding due to curing or crosslinking. The structure of the polymer may be changed irreversibly by heating. A strong covalent bond is formed between the polymer chains. Once heated, it cannot be reshaped again into the previous shape. Whereas, a thermoplastic polymer is a kind of polymer which can melt at higher temperatures and solidify at lower temperatures. It has weak intermolecular force between the polymer chains. Also, there is no crosslinking between the chains, and it can be reshaped again and again by varying the temperature. The advantages of thermoplastic polymers are lower manufacturing cost, long shelf-life, recyclability, low moisture content, *etc* [17].

Metal Matrix Composites (MMC)

Metal matrix composite is a combination of a metallic matrix (*i.e.*, aluminum, titanium, magnesium, copper, *etc*.) with reinforcement material, *i.e.,* ceramic and non-metallic phases. The desirable metallic properties of a metallic matrix are achieved through homogenization with a secondary phase resulting in high specific strength and high-specific stiffness [18]. The parameters affecting the morphologies of MMCs depend upon the properties of the matrices and reinforcement materials. Also, the interaction or adhesion of the matrix and reinforcement over the interphase determines how strong a composite would be. The interphase also helps in transferring the stress between the matrix and reinforcement. MMCs can be classified into different categories based on the matrix materials, namely:

- Aluminum-based composite.
- Magnesium-based composite.
- Titanium-based composite.
- Copper-based composite.
- Super alloy-based composite.

MMC is the most commonly used composite on the industrial scale. The advantages and disadvantages of PMC and CMC are discussed in Table 4 referred as [18]. The selection of a good metal and reinforcement material is very crucial so that a good morphology of composite can be achieved.

Table 4. Advantages and disadvantages of MMC.

Advantages	Disadvantages
• High thermal resistance. • Less moisture adsorption. • High electrical conductivity. • Higher radiation resistance. • High tensile strength and stiffness • Better abrasion resistance.	• The cost is high. • Complex synthesis methods. • Relatively immature technology.

Hence, it is essential to check the quality and nature of the metal and the reinforcement material.

Ceramic Matrix Composites (CMC)

Ceramic matrix composite is a subtype of composite in which the matrix material is ceramic and the reinforcement material is either fiber or particulate material. It has much precedence over conventional ceramics such as alumina, silicon carbide, aluminum carbide, *etc.* with respect to its morphological properties. The mechanical properties of CMC are increased by incorporating particulates or fibers in the matrix. Some of the most commonly used CMCs are C/C, C/SiC, SiC/SiC, Al_2O_3/Al_2O_3. The most recent research has shown that the incorporation of graphene in the composite increases the mechanical properties such as fracture toughening and high specific strength [19]. Moreover, it also incorporates itself with hydroxyapatite, *i.e.,* hydroxyapatite-GNP composite which is used in biomedical applications. Similarly, S. Yuan *et al.* incorporated carbon with silicon carbide matrix using a cutting force prediction dynamic model based on rotary ultrasonic machining which improved the mechanical properties of the composite such as high strength with low density, high wear resistance, high abrasion resistance, *etc* [20]. The applications of CMCs are wide in the field of aerospace, thermal protection equipment for space vehicles, rocket motors and jets engines [21]. A comparison of advantages of CMC over its disadvantages is summarized in Table **5**.

Table 5. Advantages and disadvantages of CMC.

Advantages	Disadvantages
• High thermal shock. • Creep resistance. • High wear resistance. • High specific strength and stiffness. • Corrosive and chemical resistance. • Low weight. • High temperature resistance.	• High manufacturing cost. • Low efficiency. • Hard and brittle materials.

MANUFACTURING TECHNIQUES OF COMPOSITES

Every composite material can be manufactured by different methods and is not confined to only one particular method. Sometimes, a composite can be fabricated combining two different techniques as well depending upon the nature and requirement of the composite. Hence, the composite manufacturing techniques can be classified into many types, and are chosen based on the type of the matrix and the reinforcement material. However, some of the commonly used techniques are discussed below [22], and also a pictorial representation of the various types of such techniques has been illustrated in Fig. (**6**).

Open Molding

In this technique, curing is done at the room temperature and a simple process of molding is used.

Hand Lay-up

It is the most widely and frequently used method due to its simplicity. It is also the oldest technique for preparation of the composites. A gel is coated on the open mold and then dried fiber in the form of woven, knitted, stitched, or bonded fabric is used as the fiber reinforcement [23]. Then the resin is applied externally onto the fiber matrix with a brush and the rotary rollers.

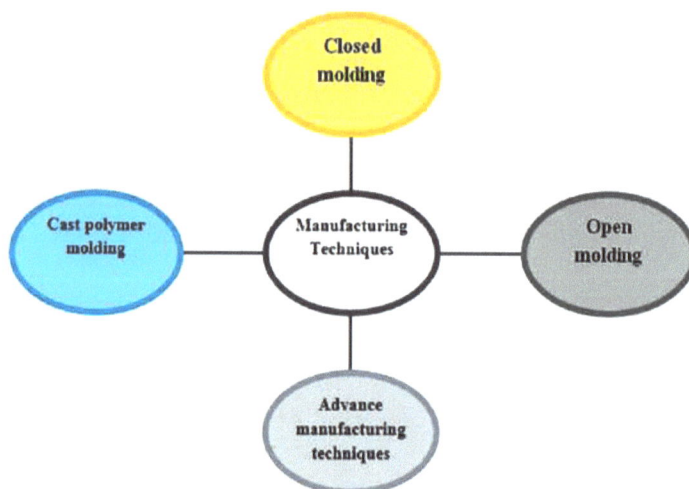

Fig. (6). Classification of composite manufacturing techniques.

Then, these are used for even distribution of the resin. Curing is accomplished at room temperature. However, to increase the speed of the process it may be heated up with the use of an oven, and reduced pressure is applied under vacuum conditions.

Spray Up

The fiber used in the spray up method is finely chopped down and is used as a reinforcement material. Resins are used as the polymer matrices, mixed together with a spray gun, and sprayed simultaneously into the mold. A separate roller is used to reinforce fiber with resin [24, 25]. Then it is left under room temperature for curing. This process is faster than the hand lay-up method because the fibers, being broken down into small size, give good conformability to the composite. The thickness and consistency of the coating can be manipulated as it also allows site fabrication.

Filament Winding

An automated open molding fabrication method using continuous roving with a rotating mandrel is used in this process. The fibers are taken from continuous roving and pass through the hot resin bath. Then the resin-infused continuous fiber is wrapped around the rotating mandrel depending upon the requirement of layers [26]. Then the laminate is left for curing at room temperature. This technique provides high fiber content up to 70% of the volume and gives good mechanical properties. It is often used in the manufacturing of open cylinders and closed ones too (pressure vessels or tanks) [27]. If the applied resin is in the liquid state and is applied on the filament, then it is known as wet filament winding; and if the resin is sprayed over the filament, then it is known as dry filament winding.

Closed Molding

The process is automated and is used for large-scale production.

Vacuum Bag Molding

It is a modification of the hand lay-up process. The benefits of using this technique are that it removes voids and humidity from the laminates. Also, it improves the fiber-to-resin ratio which is a key factor in improving the strength-to-weight ratio of the composites. The laminate processed in the hand lay-up method is put in a vacuum chamber. Then a flexible thin film is placed over the wet lay-up and the reinforcement is separated with resin. The material is packed under a vacuum bag and atmospheric pressure which helps in avoiding air voids, excess resin, and humidity [28].

Reinforced Reaction Injection Molding (RRIM)

Two or more resins are polymerized together to form a thermosetting polymer under high pressure. Glass fiber and mica are incorporated into the mixture to improve the quality of the composite. Then the mixture is poured into the mold with the help of an injection cylinder at high pressure. The stream of the mixture is mixed under high pressure, but the resultant liquid has a low viscosity that is injected into the mold cavity under low pressure about 50 psi, and polymerization takes place very quickly [22, 29].

Centrifugal Casting

It is a casting technique that uses centrifugal force for the mixing of resin and reinforcement material. The resin-reinforcement mixture is deposited in the rotating closed cylindrical mold. The mixture is under continuous rotation with centrifugal force until it gets cured. The formed molten resin is poured down into the mold cavity. It has applications in manufacturing pipes. Nano-sized TiB_2 particles with an aluminum matrix are produced with high particulate volume percentage [30, 31].

Cast Polymer Molding

Fiber reinforcement is not used in this method and it is specifically designed as per the application [22].

Gel Coated Culture Stone Molding

A redeveloped polyester resin material is used as a gel coat which provides protection to the composites externally. The gel coat is sprayed over the molding surface and left for curing. Fillers are used as reinforced materials instead of fibers for enhanced properties and certain composites [22].

Solid Surface Molding

A mixture of polyester resin or acrylic resin is used to make the surface of a material look enticing and attractive. Molding of the resin mixture is accomplished under vacuum conditions in order to avoid the voids in the matrix. An example is coating on natural granite stone [32].

Engineered Stone Molding

Engineered stone molding refers to the technique that combines resin with the stone particles in a vacuum chamber. Molding is carried out at vacuum conditions which provides an air-free matrix, and then it is compressed to achieve low porosity casting. Incorporation of the stone particles in the matrix increases strength, and provides high heat resistance, low thermal expansion, strain and scratch resistance [33].

Advanced Manufacturing Technique

It is also known as an additive manufacturing technique. The advanced manufacturing technique is a process that surpasses conventional manufacturing methods due to its advantages over the conventional methods. In conventional methods, molding is required which makes the process complex and difficult and is also not cost-effective. Hence, these lacunae are filled by using the additive manufacturing technique in which composite structures are made layer-by-layer, and also the composites can be designed easily with the aid of computers. It is processed mainly in four steps [22]:

Material Extrusion

Continuous filament thermoplastic or composite materials are deposited layer-by-layer through a nozzle. Filaments are melted at a high temperature followed by deposition. This mechanism is known as fused filament fabrication (FFF) [34].

Vat Photo-polymerization

In this method, curing of the material is done by UV light and it gives high-resolution material. It is also known as light-activated polymerization.

Sheet Lamination

The piling of sheets is done layer-by-layer to form a structure. It is of two types: laminated object manufacturing (LOM) and composite-based additive manufacturing (CBAM).

Powder Bed Fusion

It uses a laser or an electron beam to melt and fuse powder materials together.

APPLICATION OF POLYMER COMPOSITE IN TISSUE ENGINEERING

Polymer composites have a wide range of applications, *i.e.,* from manufacturing aerospace industries to biomedical industries. One of the recent developments which have received immense interest is the utilization of biomaterial in tissue engineering. A biomaterial-based scaffold is developed which acts as an external support to the tissue and gives a platform for the cell to grow. The chemical and physicochemical compositions of scaffolds play a crucial role in cell growth, proliferation, cell attachment and cytocompatibility [35]. The customization of a scaffold can be done chemically, physically and through surface modification *via* various fabrication methods. The condition of fabrication methods can depend upon the requirements of tissue and cell. The process of developing a biomaterial includes material selection for the biomaterial, and synthesis by using different methods as per requirement. In the development of a scaffold, the cells are seeded and cultured under controlled conditions which leads to the fabrication of functional tissues. It is further explained in Fig. (**7**). The addition or combination of biomaterials can be made with nanoparticles, fiber, ceramic and polymer. The incorporation of nanoparticles in a matrix forms nanocomposites. This nanocomposite has enhanced physical and chemical properties compared to the conventional composites, as described (development) in Fig. (**8**), such as large surface area, improved cell proliferation, adherence and differentiation, strength and stability of the composite [36]. The composite-based biomaterials can be classified into two categories: natural and synthetic composites. Naturally occurring composites are used due to their biocompatibility, biodegradability and availability. Examples are chitosan, collagen, alginate, cellulose, *etc*. Synthetic composites are non-immunogenic. Their mechanical properties and shape can be easily modified, and its quality is also good. Examples are polycaprolactone, poly lactic-co-glycolic, polyethylene glycol, polylactic acid, polyurethane, *etc*. This type of composite is not only used in the synthesis of scaffolds but also in drug delivery systems. The classes of such composite are discussed below.

Natural Polymers

These polymers are bioactive in the environment, similar in structure, and are compositionally used in tissue engineering applications. These are biodegradable and biocompatible in nature which is why these are being used in bone tissue engineering (BTE). Some of the commonly used natural polymers are chitosan, collagen, alginate, silk, cellulose, *etc*., and their advantages are given below in Table **6**.

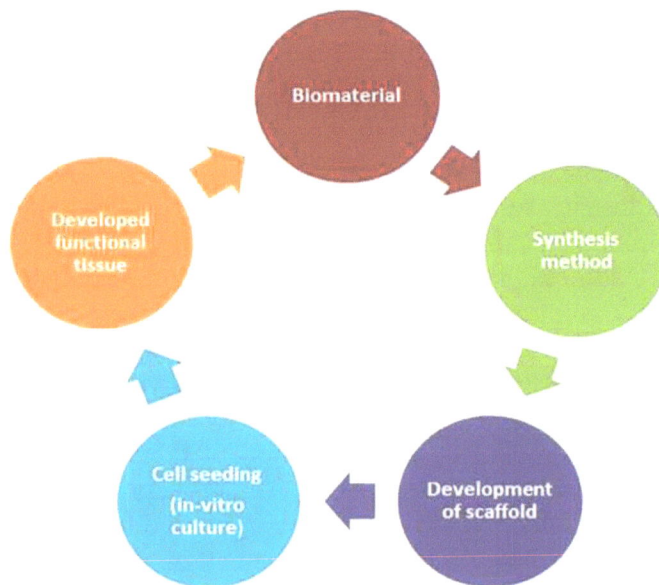

Fig. (7). Schematic representation of the development of functional tissue with scaffold.

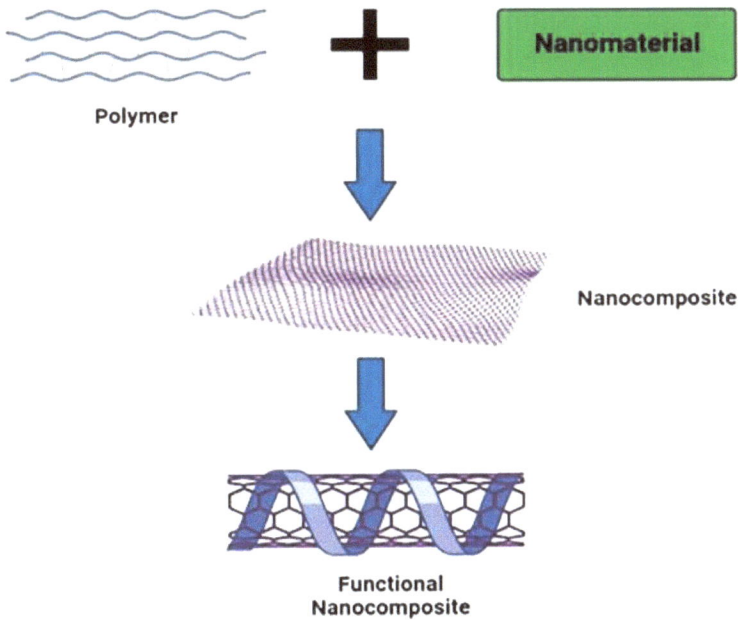

Fig. (8). Development of nanocomposites.

Table 6. Applications of different natural polymers.

Polymer Used	Synthesis Method	Nanoparticle Used	Advantage	Reference
Chitosan	• Sol-gel method and freeze-drying method. • Solution-based chemical method.	nHap	• Increased thermal and mechanical stability. • Increased cytocompatibility.	[39] [40]
Collagen	Freeze drying method.	• F-MWCNT and nHap. • BCP with dexamethasone.	• Increased porosity, mechanical strength and bioactivity. • Improved cell adhesion and proliferation.	[42] [43]
Silk fibroin	Phase separation method.	Fluoridated TiO_2.	Enhanced cell attachment, cell cytotoxicity, biocompatibility and bioactivity.	[46, 47]
Cellulose	Biomimetic approach	Hap and CNW	The biomimetic material exhibited much better physicochemical and biocompatibility properties than the wet chemical material.	[51]
Alginate	• Lyophilization method • Freeze drying method	• Gelatin/nHap • CS, nHap and fucoidan	• Better thermal stability, biodegradability and increased crystallinity. • Biomineralization and biocompatibility increased.	[54] [55]

Chitosan-based Nanocomposites

Chitosan is a naturally occurring polymer, found abundantly in nature. It is a β-1, 4-linked N acetyl D-glucosamine and D-glucosamine unit which is highly degradable in nature [37, 38]. Due to its cytocompatibility with the tissues, it is a promising biomaterial for BTE. Several studies have shown that chitosan-based (CS) scaffold does not have complications such as inflammation, post-implantation, and it is non-toxic to the tissue [38]. However, there are certain limitations to the usage of a simple CS scaffold that is why it is incorporated with particles and minerals such as hydroxyapatite (Hap), collagen, gelatin, tri-calcium phosphate, and synthetic polymers. The most commonly used nanofiller is hydroxyapatite because of its structure and similarity of composition. In a study, the incorporation of nHap with a composite matrix exhibited improved properties of the composite such as compressive strength and modulus, thermal efficiency, increased surface roughness, higher biomineralization, and adequate cell proliferation [39]. Another study conducted by N. Pramanik *et al.* was based on the combination of chitosan phosphate (CSP) with hydroxyapatite which was synthesized using the solution-based chemical method [40]. The physicochemical properties were characterized using XRD, FTIR, SEM, TEM, *etc.* It revealed that the increase in Hap content increases the mechanical properties of the nanocomposite. Cytocompatibility test was done using murine L929 fibroblast cell lines resulting in cytocompatibility with the cellular environment.

Collagen-based Nanocomposites

Collagen is found in the extracellular matrix (ECM) of bone naturally. It is used in the transport of nutrients to bone cells and also nourishes the bone cells. Collagen is a bioactive natural polymer found in the extraction of fish waste such as bones, scales, and skins. Due to collagen being found inside the ECM, it is an excellent biocompatible material for tissue engineering, and facilitates cell attachment and proliferation, and faster biodegradability, but increases the swelling of the scaffold and results in poor mechanical strength [41, 42]. These non-advantageous attributes of collagen scaffold may be improved by functionalizing the collagen composite with a nanofiller such as functionalized multiwall carbon nanotube (f-MWCNT), which can be synthesized by using collagen and hydroxyapatite (HA) using the freeze-drying technique [42]. The incorporation of f-MWCNT improved the porous structure with increased surface area. Also, an improvement in mechanical strength and biocompatibility was observed. However, the incorporation of the nanofiller decreased the porosity of the material. On the contrary, a similar study was conducted by Y. Chen *et al.* who synthesized a composite incorporated with biphasic calcium nanoparticle (BCP) along with dexamethasone using the freeze-drying method [43]. It revealed increased porosity of scaffold and biocompatibility, improved mechanical strength, and also induced excellent cell proliferation.

Silk Fibroin Nanocomposites

Silk fibroin (SF) is a naturally occurring protein that is extracted from *Bombyx mori* cocoon, a mulberry source. Silk fibroin is found in the structural protein and is a biocompatible material. It provides enhanced mechanical properties, improved cell adhesion and proliferation, and excellent cytocompatibility which are some of the escalated properties of SF composite scaffold [44, 45]. In contrast to that, a study was conducted by N. Johari *et al.* where a scaffold was fabricated using SF and fluoridated TiO_2 nanoparticles [46]. TiO_2 was fluoridated with the fluoride ions at different volume concentrations, and embedded in the SF matrix through the phase separation method. The mechanical properties, bioactivity and biocompatibility of the scaffold were improved. Moreover, the cell attachment as well as cell cytotoxicity were also enhanced. Complementarily, a study was also performed by N. Johari *et al.* on the preparation of scaffold with the same SF and TiO_2 using freeze-drying method [47]. The biocompatibility of a scaffold was investigated using SaOS-2 osteoblast-like cells. Even though mechanical strength increased, but a reduction in porosity was also observed with the increase in the nanoparticle content. The rise of nanoparticles resulted in its agglomeration on the pore walls leading to blockage of the pores and reduction of porosity. However, the clustering of the nanoparticles also increased the mechanical strength of the scaffold. Biomineralization, cell adhesion and proliferation, and cell viability were elevated in the SaOS-2 osteoblast-like cells with increased SF and TiO_2 contents [47].

Cellulose-based Nanocomposites

Cellulose is a polysaccharide compound, having a linear chain of β-1,4-linked glucose units. It is found in plants, bacteria and algae. The characteristics of cellulose such as biocompatibility, biodegradability, non-toxic character, presence of protein binding sites, excellent *in-vivo* stability, and amendable properties make it a suitable biomaterial for BTE and drug delivery as well [48]. Since it has certain limitations such as, no definite 3D structure and low mechanical properties; therefore, it needs to be coupled with other natural and synthetic materials to overcome its limitations [49, 50]. A comparative study between the biomimetic method and the wet chemical precipitation method was evaluated [51]. The biomimetic approach has been extensively used for the synthesis of cellulose-based nanocomposites for tissue engineering applications. The growth of Hap was reinforced on a cellulose nanowhisker (CNW) scaffold. CNW was synthesized using hydrochloric acid, phosphoric acid and sulfuric acid. Moreover,

Hap was also synthesized by using simulated body fluid (SMF) at 1.5 M concentration. The nucleation potential and the growth of Hap were directly impacted by the sulfonate and phosphonate groups of CNW. Fibrous cell (L929) was used for assessing the bioactivity and biocompatibility of the hybrid material by cell viability test. The result obtained showed that the biomimetic approach had an upper hand and more advantages compared with the wet chemical approach scaffold [51].

Alginate-based Nanocomposites

Alginate is a naturally occurring edible polysaccharide, obtained from brown algae such as *Laminaria japonica* and *Macrocystis pyrifera* [52]. *Pseudomonas* and *Azotobacter* are the bacteria that also produce alginate [53]. Alginate has great biocompatibility which makes it an excellent biomaterial for BTE. In addition to that, it also allows uniform distribution of growth factor, easy ligand attachment and size handling, and conformance with irregular-shaped bone defects. Having said that, still it has poor mechanical and tensile strength. Its fabrication with nanoparticles and polymers can improve these properties. Alginate-di-aldehyde (ADA), reinforced with cross-linked gelatin/nano-hydroxyapatite (nHap) bio-scaffold, was synthesized using the lyophilization method [54]. The functional groups of ADA and gelatin reacted, *i.e.,* crosslinking between aldehyde in ADA and the amino functional group of gelatin was formed which helped in the improvement of thermal stability of the scaffold. Moreover, the addition of nHap aided in improving the scaffold stiffness, biodegradability and crystallinity, but decreased its porosity. In another study, a 3D scaffold was synthesized using Chitosan, nHap, and fucoidan with the technique, called the freeze-drying method [55]. A reduction in water absorption and retention was exhibited. However, the porosity of the scaffold did not decrease with the increasing amount of nanofiller; whereas, biomineralization of the scaffold increased. Biocompatibility test was done with periosteum mesenchymal stem cells (PMSC) which revealed good cell differentiation, proliferation, growth and mineral deposition [55].

Synthetic Polymers

Long molecular chain polymers, artificially synthesized from small monomers, are known as synthetic polymers. These have advantages over natural polymers due the flexibility in their properties. Synthetic polymers are also used widely in bone tissue engineering application and drug delivery. Some of the commonly used synthetic polymers are polycaprolactone (PCL), poly D,L-lactic-co-glycolic acid (PLGA), polylactic acid (PLA), polyethylene glycol (PEG) *etc*. In Table **7,** various advantages of different synthetic polymers have been discussed.

Table 7. Application of synthetic polymers.

Polymer Used	Synthesis Method	Nanoparticle Used	Advantages	References
PCL	*In-situ* solvothermal method.	Hap nanorods.	• Superior elastic modulus and strength. • Enhanced cell adhesion and proliferation.	[57] [58]
PLGA	• Electrospinning method. • Sintering method.	• SF and GO. • TNT	• Enhanced young modulus and tensile strength, increased cell adhesion and proliferation. • 3D porous scaffold with increased bioactivity and cell viability.	[61] [62]
PEG	• Electrospinning method. • Flame spray pyrolysis.	• PCL nanofiber and GO. • nCP	• Increased fiber diameter, young modulus and mechanical strength. • Increase in water absorption and biomineralization, but a reduction in the roughness of nanocomposite.	[63] [64]
PLA	• Phase separation technique. • Solvent casting and salt-leaching method.	• Chitosan –Hap hydrogel. • nHap	• Improvement in growth and differentiation of cells. Increased pore size and cell viability. • Increased nucleation site increases biomineralization and enhances cell adhesion, proliferation, and cell distribution.	[67] [70]
PU	Electrospinning method	• $nTiO_2$ and PEUU • fMWCNT and ZnO nanoparticle	• Fiber diameter, tensile stress and modulus of elasticity increased. • Enhancement in biocompatibility, biodegradability, biomineralization and mechanical strength.	[73] [74]

Polycaprolactone-based Nanocomposites

PCL is an aliphatic biodegradable polyester having exceptional mechanical strength and firmness. Biodegradation of PCL occurs by hydrolysis method, but the process of degradation is very slow; and because of this, it poses a negative impact on the tissue engineering process [56]. Moreover, the hydrophobic nature of PCL decreases cell adhesion and proliferation which can be amended by the incorporation of inorganic compounds such as titanium dioxide, hydroxyapatite, *etc* [57]. A solvothermal method was used to prepare PCL nanocomposite reinforced with Hydroxyapatite nanorods under *in-situ* conditions [58]. The precursor and the polymer solution were synthesized in an autoclave at different temperatures ranging from 60-150 ^{0}C. A comparison was made between the *in-situ* and *ex-situ* processed scaffolds. The *in-situ* processed nanocomposite showed a highly porous scaffold ($\geq$90%) containing Hap nanorods and it possessed superior elastic modulus and strength than the *ex-situ* nanocomposite. MTT assay was also conducted to analyze the bioactivity of the scaffold which also showed enhanced cell adhesion and proliferation of human osteosarcoma cell lines [58].

Poly (Lactic-co-Glycolic) Acid-based Nanocomposites

PLGA is a synthetic biodegradable polymer composed of a dimer of glycolic acid and lactic acid. Biodegradability, biocompatibility, and controlled degradation rate are the properties of PLGA that make it an outstanding polymer component for BTE scaffold [59]. In contrast to that, lack of adequate mechanical strength leads to poor mechanical characteristics of the nanocomposites which can be fixed by incorporating several ceramic nanoparticles like flourohydroxyapatite, nHap and tri-calcium phosphate [60]. In a study conducted, a blend of PLGA, tussah silk fibroin, and graphene oxide (GO) was fabricated using the electrospinning method [61]. Upon blending, it increased the fiber diameter of the nanocomposite comprising 1wt% GO and 10wt% SF in PLGA. The formation of a hydrogen bond between the functional group of the polymer matrix and nanofiller enhanced the young modulus and tensile strength of the scaffold. Furthermore, a significant enhancement in cell attachment, proliferation, and biomineralization was also observed. Similarly, in another literature, a 3D porous scaffold made of PLGA and TNT (TiO$_2$ nanotube) was synthesized by the sintering method [62]. The mechanical strength and pore structure of the scaffold were analyzed after sintering at 100 ^{0}C for 3h. Moreover, the increased bioactivity and cell viability were observed using MTT assay and alkaline phosphatase activity. The stress distribution of the PLGA/TNT scaffold over bone was reduced and it was evaluated by the 3D finite element model.

Polyethylene Glycol-based Nanocomposites

PEG is a hydrophilic polyether and is also biocompatible in nature. It has a wide range of applications, from being used as additives in cosmetics and foods, to composite materials for biomedical applications. A PCL nanofiber scaffold with different concentrations of graphene oxide (GO) and modified GO (surface treated with PEG) was synthesized using the electro spinning method [63]. The addition of PEG-grafted GO to the PCL scaffold increases the viscosity of the solution which increases the fiber diameter, young modulus and hydrophilicity. Uniform distribution of nanoparticle was also obtained from GO/PEG grafting which may be attributed as a reason for increased mechanical properties of the scaffold. The cytocompatibility of a scaffold such as cell attachment and growth was analyzed with MC3T3- E1 cells using MTT assay. In another study, PCL/PEG/SF-based membrane, incorporated with nano calcium phosphate, was developed for guided bone regeneration (GBR) [64]. Nano-calcium phosphate was synthesized using the flame spray pyrolysis (FSP) technique. The incorporation of nCP increased the diameter of fiber and also the mechanical strength of the composite. Increased water absorption, enhanced biomineralization, and reduced roughness were observed in the membrane with increased nCP. The high content of nCP also induced cell adhesion and proliferation by osteocompatibility study.

Poly (Lactic Acid)-based Nanocomposites

PLA is a synthetic polymer used for the fabrication of a scaffold in tissue engineering applications because of its biodegradability and biocompatibility properties [65]. FDA has also approved PLA as non-toxic and safe for usage in biomedical applications. Tissue engineering (TE) requires a scaffold that has excellent mechanical properties; but PLA lacks adequate mechanical strength, because of which it undergoes certain modifications and fabrication such as surface modification, blending with a polymer fiber and incorporation of nanoparticles [66]. A combination of 3D printing method and phase separation technique was used to fabricate PLA with chitosan-hydroxyapatite porous hydrogel [67]. The interaction between the carbonyl and the amino acid of CS and PLA polymer significantly increased the pore size by around 60%. An increase in the pore size of the scaffold facilitates growth and differentiation of human mesenchymal stem cells which distribute the nutrients uniformly and result in better cell survival across the scaffold. Moreover, low water retention ability and reduced elastic modulus in the wet state resulted due to Hap nanoparticles [67 - 69]. In addition to that, another study was conducted, where a combination of solvent casting and salt leaching technique was used for the generation of Hap/PLA scaffold [70]. The FESEM study revealed that the nanoparticles were dispersed uniformly around the scaffold, and thus leading to a reduction in the pore size. However, an increase in Hap nanoparticles induced nucleation site in the scaffold, and thus enhanced biomineralization potential of the scaffold. Enhancement in cell distribution, cell proliferation, and cell adhesion were also observed using a test with MG63 osteoblast cell lines compared to the scaffold having a lower percentage of Hap. However, an increase in the Hap content leads to a reduction of cell viability [70].

Polyurethane-based Nanocomposites

Polyurethane (PU) is a polymer composed of organic units joined by carbamate links. It is composed of two segments, *i.e.,* hard segment (made of diisocyanates and diamines) and soft segment (made of macro-diols), also known as chain extender diols [71]. PU has favorable properties for TE such as biocompatibility, mechanical and physical attributes [72]. Composites constructed with PU have enhanced mechanical properties while original toughness is reduced. So, they have to undergo certain modifications such as the incorporation of nanoparticles to enhance their surface area to volume ratio. A $nTiO_2$ grafted with PU was used as a reinforcement in fibrous and degradable poly(ester urethane)-urea (PEUU). This process improved interfacial bonding of $nTiO_2$ and PEUU [73]. An increase in the mechanical properties of PEUU scaffold, such as tensile stress and modulus of elasticity was observed in case of grafted $nTiO_2$ in contrast to the non-grafted $nTiO_2$. In addition to that, the increase in fiber diameter, uniform distribution of particle, and strong interaction between PEUU polymer matrix and grafted $nTiO_2$ were observed in grafted $nTiO_2$ as compared to non-grafted $nTiO_2$. Cytocompatibility analysis was also performed with bone marrow-derived mesenchymal stem cells which showed the highest cell proliferation with grafted $nTiO_2$ while compared to the controlled sample. Moreover, increased biomineralization activity was also observed. Similarly, in another study, a functionalized multi-wall carbon nanotube (FMWCNT) and ZnO nanoparticle combined with PU fiber were used for the generation of a scaffold for BTE [74]. By TEM data it was concluded that the biocompatibility, biodegradability, mineralization, and mechanical strength were improved due to the uniform dispersion of FMWCNT and nZnO over the scaffold. Moreover, biomineralization increased with an increase in the nucleation rate of the calcium and phosphate ions due to high negatively charged groups on the nanofillers elevating the adsorption properties of the calcium and phosphate ions. Cytocompatibility analysis was performed with MC3T3-E1 (preosteoblast) cells for osteoblast differentiation.

CONCLUDING REMARKS

Composite materials have superior physicochemical properties compared to conventionally used materials. High strength with less weight, high thermal resistance, better abrasion resistance, wear and tear resistance are some of the characteristics of composite materials. The use of nanocomposites in tissue engineering has increased over the years due to their potential in developing functional tissue that could replace the damaged tissue. It has several advantages such as biodegradability, biocompatibility, bioactivity and non-toxic nature. However, nanocomposites also have several limitations such as, there is a need to enhance the function of any composite material having nanofiller or external particles. Increase in usage of nanofiller or nanoparticle sometimes decreases the conventional properties of the composites. Surface modification, functionalization and grafting with other particles might increase some of the properties, but however, it also leads to a decrease in some of the other properties of the composites. Having said that in future, a researcher must find a way in which the incorporation of nanofiller should be done resulting in minimal damage; or find a naturally occurring material which doesn't require the incorporation of external particles. Application of the composites in tissue/bioengineering level is another challenge that has to be countered. However, certain advancements and great potentials of polymer composites have been shown by various reports and research works done by people over the years. However, the exact mechanism of cytotoxicity of the composites, the precise mechanism of the composite materials targeting the cell, would the dissolution of composites take place intracellularly or extracellularly, and what would be the impact on the cell after several years of exposure are some of the factors which we should take into consideration, and future studies can be done on these parameters.

ACKNOWLEDGEMENTS

The authors are grateful to the Council of Scientific and Industrial Research (CSIR), New Delhi, India for providing financial support (Project grant no. 22(0847)/20/EMR-II, dated: 10.12.2020). The authors also gratefully acknowledge to National Institute of Technology (NIT), Arunachal Pradesh, India, for assistance and support.

REFERENCES

[1] Wang X, Jiang M, Zhou Z, Gou J, Hui D. 3D printing of polymer matrix composites: A review and prospective. Compos, Part B Eng 2017; 110: 442-58.
[http://dx.doi.org/10.1016/j.compositesb.2016.11.034]

[2] Naskar AK, Keum JK, Boeman RG. Polymer matrix nanocomposites for automotive structural components. Nat Nanotechnol 2016; 11(12): 1026-30.
[http://dx.doi.org/10.1038/nnano.2016.262] [PMID: 27920443]

[3] Wang RM, Zheng SR, Zheng YP. Polymer matrix composites and technology. Elsevier 2011; pp. 1-548.
[http://dx.doi.org/10.1533/9780857092229]

[4] Saheb DN, Jog JP. Natural fiber polymer composites: A review. Adv Polym Technol 1999; 18(4): 351-63.
[http://dx.doi.org/10.1002/(SICI)1098-2329(199924)18:4<351::AID-ADV6>3.0.CO;2-X]

[5] Giliopoulos D, Zamboulis A, Giannakoudakis D, Bikiaris D, Triantafyllidis K. Polymer/metal organic framework (MOF) nanocomposites for biomedical applications. Molecules 2020; 25(1): 185.
[http://dx.doi.org/10.3390/molecules25010185] [PMID: 31906398]

[6] Prashanth S, Subbaya KM, Nithin K, Sachhidananda S. Fiber reinforced composites-a review. J Mar Sci Eng 2017; 6(3): 2-6.

[7] Lotfi A, Li H, Dao DV, Prusty G. Natural fiber–reinforced composites: A review on material, manufacturing, and machinability. J Ther Comp Mat 2021; 34(2): 238-84.
[http://dx.doi.org/10.1177/0892705719844546]

[8] Miracle DB, Donaldson SL, Henry SD, *et al.* ASM handbookMaterials Park, OH: ASM international 2001; 21: pp. 21-57.

[9] Yang G, Park M, Park SJ. Recent progresses of fabrication and characterization of fibers-reinforced composites: A review. Comp Comm 2019; 14: 34-42.
[http://dx.doi.org/10.1016/j.coco.2019.05.004]

[10]   Campbell FC. Structural composite materials. ASM international 2010.
       [http://dx.doi.org/10.31399/asm.tb.scm.9781627083140]

[11]   Zhang G, Lu H, Mamidwar S, Wang M. Composite, Biomater Sci. An introduction to materials in
       medicine. 2020; 4: 415-29.

[12]   Qing Y, Min D, Zhou Y, Luo F, Zhou W. Graphene nanosheet- and flake carbonyl iron particle-filled
       epoxy–silicone composites as thin–thickness and wide-bandwidth microwave absorber. Carbon 2015;
       86: 98-107.
       [http://dx.doi.org/10.1016/j.carbon.2015.01.002]

[13]   Khayal OMES. Advancements in polymer composite structure. 2019; pp. 1-15.

[14]   Bednarcyk BA. An inelastic micro/macro theory for hybrid smart/metal composites. Compos, Part B
       Eng 2003; 34(2): 175-97.
       [http://dx.doi.org/10.1016/S1359-8368(02)00067-7]

[15]   Huang H, Talreja R. Numerical simulation of matrix micro-cracking in short fiber reinforced polymer
       composites: Initiation and propagation. Compos Sci Technol 2006; 66(15): 2743-57.
       [http://dx.doi.org/10.1016/j.compscitech.2006.03.013]

[16]   Pelegrín YF, Santana TJM. Characterization techniques for algae-based materials.In algae based
       polymers, blends, and composites. Elsevier 2017; pp. 649-70.
       [http://dx.doi.org/10.1016/B978-0-12-812360-7.00018-5]

[17]   Odian G. Principles of polymerization. John Wiley & Sons 2004.
       [http://dx.doi.org/10.1002/047147875X]

[18]   Haghshenas M. Metal–matrix composites, Reference module in materials science and materials
       engineering. Elsevier 2016.

[19]   Nieto A, Bisht A, Lahiri D, Zhang C, Agarwal A. Graphene reinforced metal and ceramic matrix
       composites: a review. Int Mater Rev 2017; 62(5): 241-302.
       [http://dx.doi.org/10.1080/09506608.2016.1219481]

[20]   Yuan S, Fan H, Amin M, Zhang C, Guo M. A cutting force prediction dynamic model for side milling
       of ceramic matrix composites C/SiC based on rotary ultrasonic machining. Int J Adv Manuf Technol
       2015; 86(1): 37-48.

[21]   Krenkel W, Berndt F. C/C–SiC composites for space applications and advanced friction systems.
       Mater Sci Eng A 2005; 412(1-2): 177-81.
       [http://dx.doi.org/10.1016/j.msea.2005.08.204]

[22]   Rajak DK, Pagar DD, Kumar R, Pruncu CI. Recent progress of reinforcement materials: a
       comprehensive overview of composite materials. J Mater Res Technol 2019; 8(6): 6354-74.
       [http://dx.doi.org/10.1016/j.jmrt.2019.09.068]

[23]   Davim JP, Reis P, António CC. Experimental study of drilling glass fiber reinforced plastics (GFRP)
       manufactured by hand lay-up. Compos Sci Technol 2004; 64(2): 289-97.
       [http://dx.doi.org/10.1016/S0266-3538(03)00253-7]

[24]   Karlsson KF, Tomas Åström B. Manufacturing and applications of structural sandwich components.
       Compos. Part. Appl Sci (Basel) 1997; 28(2): 97-111.

[25]   Campbell FC, Chapter 11, Commercial composite processes: These commercial processes produce far
       more parts than the high-performance processes, manufacturing processes for advanced composites,
       Elsevier 2004; 399-438.

[26]   Mazumda S. Composites manufacturing: materials, product, and process engineering. CRC Press
       2001.
       [http://dx.doi.org/10.1201/9781420041989]

[27]   Advani SG, Hsiao KT. Manufacturing techniques for polymer matrix composites (PMCs). Elsevier

2012.
[http://dx.doi.org/10.1533/9780857096258]

[28]   Goren A, Atas C. Manufacturing of polymer matrix composites using vacuum assisted resin infusion molding. Arch Mater Sci Eng 2008; 34(2): 117-20.

[29]   Wittemann F, Maertens R, Bernath A, Hohberg M, Kärger L, Henning F. Simulation of reinforced reactive injection molding with the finite vol. method. Journal of Composites Science 2018; 2(1): 5.
[http://dx.doi.org/10.3390/jcs2010005]

[30]   Sánchez M, Rams J, Ureña A. Fabrication of aluminium composites reinforced with carbon fibres by a centrifugal infiltration process. Compos, Part A Appl Sci Manuf 2010; 41(11): 1605-11.
[http://dx.doi.org/10.1016/j.compositesa.2010.07.014]

[31]   El-Galy IM, Ahmed MH, Bassiouny BI. Characterization of functionally graded Al-SiC p metal matrix composites manufactured by centrifugal casting. Alex Eng J 2017; 56(4): 371-81.
[http://dx.doi.org/10.1016/j.aej.2017.03.009]

[32]   Bera P, Guptha N, Dasan KP, Natarajan R. Recent developments in synthetic marble processing. Rev Adv Mater Sci 2012; 32(2): 94-105.

[33]   Hamoush S, Abu-Lebdeh T, Picornell M, Amer S. Development of sustainable engineered stone cladding for toughness, durability, and energy conservation. Constr Build Mater 2011; 25(10): 4006-16.
[http://dx.doi.org/10.1016/j.conbuildmat.2011.04.035]

[34]   Turner BN, Gold SA. A review of melt extrusion additive manufacturing processes: II Materials, dimensional accuracy, and surface roughness. Rapid Prototyp. J. 2015.

[35]   Jammalamadaka U, Tappa K. Recent advances in biomaterials for 3D printing and tissue engineering. J Funct Biomater 2018; 9(1): 22.
[http://dx.doi.org/10.3390/jfb9010022] [PMID: 29494503]

[36]   Bharadwaz A, Jayasuriya AC. Recent trends in the application of widely used natural and synthetic polymer nanocomposites in bone tissue regeneration. Mater Sci Eng C 2020; 110: 110698.
[http://dx.doi.org/10.1016/j.msec.2020.110698] [PMID: 32204012]

[37]   Kavya KC, Jayakumar R, Nair S, Chennazhi KP. Fabrication and characterization of chitosan/gelatin/nSiO$_2$ composite scaffold for bone tissue engineering. Int J Biol Macromol 2013; 59: 255-63.
[http://dx.doi.org/10.1016/j.ijbiomac.2013.04.023] [PMID: 23591473]

[38]   Keller L, Regiel-Futyra A, Gimeno M, *et al.* Chitosan-based nanocomposites for the repair of bone defects. Nanomedicine 2017; 13(7): 2231-40.
[http://dx.doi.org/10.1016/j.nano.2017.06.007] [PMID: 28647591]

[39]   Das A, Ringu T, Ghosh S, Pramanik N. A comprehensive review on recent advances in preparation, physicochemical characterization, and bioengineering applications of biopolymers. Polym Bull 2022; 1-66.
[http://dx.doi.org/10.1007/s00289-022-04443-4] [PMID: 36043186]

[40]   Pramanik N, Mishra D, Banerjee I, Maiti TK, Bhargava P, Pramanik P. Chemical synthesis, characterization, and biocompatibility study of hydroxyapatite/chitosan phosphate nanocomposite for bone tissue engineering applications. Int J Biomater 2009; 2009: 1-8.
[http://dx.doi.org/10.1155/2009/512417] [PMID: 20130797]

[41]   Park JE, Park IS, Neupane MP, Bae TS, Lee MH. Effects of a carbon nanotube-collagen coating on a titanium surface on osteoblast growth. Appl Surf Sci 2014; 292: 828-36.
[http://dx.doi.org/10.1016/j.apsusc.2013.12.058]

[42]   Türk S, Altınsoy I, Çelebi Efe G, İpek M, Özacar M, Bindal C. 3D porous collagen/functionalized multiwalled carbon nanotube/chitosan/hydroxyapatite composite scaffolds for bone tissue engineering. Mater Sci Eng C 2018; 92: 757-68.

[http://dx.doi.org/10.1016/j.msec.2018.07.020] [PMID: 30184804]

[43] Chen Y, Kawazoe N, Chen G. Preparation of dexamethasone-loaded biphasic calcium phosphate nanoparticles/collagen porous composite scaffolds for bone tissue engineering. Acta Biomater 2018; 67: 341-53.
[http://dx.doi.org/10.1016/j.actbio.2017.12.004] [PMID: 29242161]

[44] Yan LP, Oliveira JM, Oliveira AL, Caridade SG, Mano JF, Reis RL. Macro/microporous silk fibroin scaffolds with potential for articular cartilage and meniscus tissue engineering applications. Acta Biomater 2012; 8(1): 289-301.
[http://dx.doi.org/10.1016/j.actbio.2011.09.037] [PMID: 22019518]

[45] Yan S, Zhang Q, Wang J, *et al.* Silk fibroin/chondroitin sulfate/hyaluronic acid ternary scaffolds for dermal tissue reconstruction. Acta Biomater 2013; 9(6): 6771-82.
[http://dx.doi.org/10.1016/j.actbio.2013.02.016] [PMID: 23419553]

[46] Johari N, Madaah Hosseini HR, Samadikuchaksaraei A. Novel fluoridated silk fibroin/ TiO_2 nanocomposite scaffolds for bone tissue engineering. Mater Sci Eng C 2018; 82: 265-76.
[http://dx.doi.org/10.1016/j.msec.2017.09.001] [PMID: 29025657]

[47] Johari N, Madaah Hosseini HR, Samadikuchaksaraei A. Optimized composition of nanocomposite scaffolds formed from silk fibroin and nano-TiO_2 for bone tissue engineering. Mater Sci Eng C 2017; 79: 783-92.
[http://dx.doi.org/10.1016/j.msec.2017.05.105] [PMID: 28629081]

[48] Saber-Samandari S, Saber-Samandari S, Gazi M, Cebeci FÇ, Talasaz E. Synthesis, characterization and application of cellulose based nano-biocomposite hydrogels. J Macromol Sci Part A Pure Appl Chem 2013; 50(11): 1133-41.
[http://dx.doi.org/10.1080/10601325.2013.829362]

[49] Beladi F, Saber-Samandari S, Saber-Samandari S. Cellular compatibility of nanocomposite scaffolds based on hydroxyapatite entrapped in cellulose network for bone repair. Mater Sci Eng C 2017; 75: 385-92.
[http://dx.doi.org/10.1016/j.msec.2017.02.040] [PMID: 28415476]

[50] Tsioptsias C, Tsivintzelis I, Papadopoulou L, Panayiotou C. A novel method for producing tissue engineering scaffolds from chitin, chitin–hydroxyapatite, and cellulose. Mater Sci Eng C 2009; 29(1): 159-64.
[http://dx.doi.org/10.1016/j.msec.2008.06.003]

[51] Fragal EH, Cellet TSP, Fragal VH, *et al.* Hybrid materials for bone tissue engineering from biomimetic growth of hydroxiapatite on cellulose nanowhiskers. Carbohydr Polym 2016; 152: 734-46.
[http://dx.doi.org/10.1016/j.carbpol.2016.07.063] [PMID: 27516325]

[52] Lee KY, Mooney DJ. Alginate: Properties and biomedical applications. Prog Polym Sci 2012; 37(1): 106-26.
[http://dx.doi.org/10.1016/j.progpolymsci.2011.06.003] [PMID: 22125349]

[53] Remminghorst U, Rehm BHA. Bacterial alginates: from biosynthesis to applications. Biotechnol Lett 2006; 28(21): 1701-12.
[http://dx.doi.org/10.1007/s10529-006-9156-x] [PMID: 16912921]

[54] Mehedi Hasan M, Nuruzzaman Khan M, Haque P, Rahman MM. Novel alginate-di-aldehyde cross-linked gelatin/nano-hydroxyapatite bioscaffolds for soft tissue regeneration. Int J Biol Macromol 2018; 117: 1110-7.
[http://dx.doi.org/10.1016/j.ijbiomac.2018.06.020] [PMID: 29885393]

[55] Lowe B, Venkatesan J, Anil S, Shim MS, Kim SK. Preparation and characterization of chitosan-natural nano hydroxyapatite-fucoidan nanocomposites for bone tissue engineering. Int J Biol Macromol 2016; 93(Pt B): 1479-87.
[http://dx.doi.org/10.1016/j.ijbiomac.2016.02.054] [PMID: 26921504]

[56] Karuppuswamy P, Reddy Venugopal J, Navaneethan B, Luwang Laiva A, Ramakrishna S. Polycaprolactone nanofibers for the controlled release of tetracycline hydrochloride. Mater Lett 2015; 141: 180-6.
[http://dx.doi.org/10.1016/j.matlet.2014.11.044]

[57] Fabbri P, Bondioli F, Messori M, Bartoli C, Dinucci D, Chiellini F. Porous scaffolds of polycaprolactone reinforced with *in situ* generated hydroxyapatite for bone tissue engineering. J Mater Sci Mater Med 2010; 21(1): 343-51.
[http://dx.doi.org/10.1007/s10856-009-3839-5] [PMID: 19653069]

[58] Moeini S, Mohammadi MR, Simchi A. *In-situ* solvothermal processing of polycaprolactone/ hydroxyapatite nanocomposites with enhanced mechanical and biological performance for bone tissue engineering. Bioact Mater 2017; 2(3): 146-55.
[http://dx.doi.org/10.1016/j.bioactmat.2017.04.004] [PMID: 29744424]

[59] Tahriri M, Moztarzadeh F. Preparation, characterization, and *in vitro* biological evaluation of PLGA/nano-fluorohydroxyapatite (FHA) microsphere-sintered scaffolds for biomedical applications. Appl Biochem Biotechnol 2014; 172(5): 2465-79.
[http://dx.doi.org/10.1007/s12010-013-0696-y] [PMID: 24395697]

[60] Eslami H, Solati-Hashjin M, Tahriri M. The comparison of powder characteristics and physicochemical, mechanical and biological properties between nanostructure ceramics of hydroxyapatite and fluoridated hydroxyapatite. Mater Sci Eng C 2009; 29(4): 1387-98.
[http://dx.doi.org/10.1016/j.msec.2008.10.033]

[61] Eslami H, Azimi Lisar H, Jafarzadeh Kashi TS, *et al.* Poly(lactic-co-glycolic acid)(PLGA)/TiO $_2$ nanotube bioactive composite as a novel scaffold for bone tissue engineering: *In vitro* and *in vivo* studies. Biologicals 2018; 53: 51-62.
[http://dx.doi.org/10.1016/j.biologicals.2018.02.004] [PMID: 29503205]

[62] Shao W, He J, Sang F, *et al.* Enhanced bone formation in electrospun poly(l-lactic-co-glycolic acid)–tussah silk fibroin ultrafine nanofiber scaffolds incorporated with graphene oxide. Mater Sci Eng C 2016; 62: 823-34.
[http://dx.doi.org/10.1016/j.msec.2016.01.078] [PMID: 26952489]

[63] Scaffaro R, Lopresti F, Maio A, Botta L, Rigogliuso S, Ghersi G. Electrospun PCL/GO-g-PEG structures: Processing-morphology-properties relationships. Compos. Part. Appl Sci (Basel) 2017; 92: 97-107.

[64] Türkkan S, Pazarçeviren AE, Keskin D, Machin NE, Duygulu Ö, Tezcaner A. Nanosized CaP-silk fibroin-PCL-PEG-PCL/PCL based bilayer membranes for guided bone regeneration. Mater Sci Eng C 2017; 80: 484-93.
[http://dx.doi.org/10.1016/j.msec.2017.06.016] [PMID: 28866191]

[65] Ghosh S, Ghosh S, Jana SK, Pramanik N. Biomedical application of doxorubicin coated hydroxyapatite (HAp) – poly (lactide-co-glycolide) (PLGA) nancomposite for controlling osteosarcoma therapeutics. J Nanosci Nanotechnol 2020; 20(7): 3994-4004.
[http://dx.doi.org/10.1166/jnn.2020.17689] [PMID: 31968413]

[66] Li H, Qiao T, Song P, *et al.* Star-shaped PCL/PLLA blended fiber membrane *via* electrospinning. J Biomater Sci Polym Ed 2015; 26(7): 420-32.
[http://dx.doi.org/10.1080/09205063.2015.1015865] [PMID: 25671790]

[67] Ghosh S, Ghosh S, Pramanik N. Bio-evaluation of doxorubicin (DOX)-incorporated hydroxyapatite (HAp)-chitosan (CS) nanocomposite triggered on osteosarcoma cells. Adv Compos Hybrid Mater 2020; 3(3): 303-14.
[http://dx.doi.org/10.1007/s42114-020-00154-4]

[68] Ghosh S, Raju RSK, Ghosh N, Chaudhury K. Ghosh S, Banerjee I, Pramanik N. Development and physicochemical characterization of doxorubicin (DOX) encapsulated hydroxyapatite-polyvinyl alcohol (DOX-HAp-PVA) nanocomposite for repair of osteosarcoma affected bone tissues. C R Chim

2019; 22: 46-57.
[http://dx.doi.org/10.1016/j.crci.2018.10.005]

[69]   Ghosh S, Ghosh S, Atta AK, Pramanik N. A succinct overview of hydroxyapatite based nanocomposite biomaterials: fabrications, physicochemical properties and some relevant biomedical applications. J Bionanosci 2018; 12: 143-58.
[http://dx.doi.org/10.1166/jbns.2018.1515]

[70]   Nga NK, Hoai TT, Viet PH. Biomimetic scaffolds based on hydroxyapatite nanorod/poly(d,l) lactic acid with their corresponding apatite-forming capability and biocompatibility for bone-tissue engineering. Colloids Surf B Biointerfaces 2015; 128: 506-14.
[http://dx.doi.org/10.1016/j.colsurfb.2015.03.001] [PMID: 25791418]

[71]   Shahrousvand M, Mir Mohamad Sadeghi G, Salimi A. Artificial extracellular matrix for biomedical applications: biocompatible and biodegradable poly (tetramethylene ether) glycol/poly (ε-caprolactone diol)-based polyurethanes. J Biomater Sci Polym Ed 2016; 27(17): 1712-28.
[http://dx.doi.org/10.1080/09205063.2016.1231436] [PMID: 27589493]

[72]   Zdrahala RJ, Zdrahala IJ. Biomedical applications of polyurethanes: a review of past promises, present realities, and a vibrant future. J Biomater Appl 1999; 14(1): 67-90.
[http://dx.doi.org/10.1177/088532829901400104] [PMID: 10405885]

[73]   Zhu Q, Li X, Fan Z, *et al.* Biomimetic polyurethane/TiO$_2$ nanocomposite scaffolds capable of promoting biomineralization and mesenchymal stem cell proliferation. Mater Sci Eng C 2018; 85: 79-87.
[http://dx.doi.org/10.1016/j.msec.2017.12.008] [PMID: 29407160]

[74]   Shrestha BK, Shrestha S, Tiwari AP, *et al.* Bio-inspired hybrid scaffold of zinc oxide-functionalized multi-wall carbon nanotubes reinforced polyurethane nanofibers for bone tissue engineering. Mater Des 2017; 133: 69-81.
[http://dx.doi.org/10.1016/j.matdes.2017.07.049]

CHAPTER 4

Polymer Composites for Energy Storage Application

Rupesh Rohan[1,*]

[1] *Indian Rubber Manufacturers Research Association (IRMRA), Sri City Trade Centre, Sri City, District: Chittoor, Andhara Pradesh, India*

Abstract: The chapter discusses the role and application of polymers (polymers and composites) in energy storage devices. Lithium-ion batteries and supercapacitors are the two main energy storage intermittents. The chapter underscores the utilization of polymers in various roles in these devices and their effect on performance, in addition to related future aspects and expectations.

Keywords: Energy, Electrolyte, Lithium ion battery, Polymer, Supercapacitor.

INTRODUCTION

In the past two- or three decades, energy demand increased tremendously because of the rising population and their unlimited demands. Energy is needed in every form to provide the basic needs of population. The demand of energy keeps on increasing exponentially, so there is an urgent need to search for a sustainable energy resource because traditional means will not going to fulfil the rising demands [1]. Combustion of natural gas, oil and coal contributes to fulfil around 80% of the energy demand and with the rigorous use, even fossil fuels are also in depletion. Though fossil fuels are renewable but they take a high amount of time so alternate sources of energy such as solar energy through UV rays, tidal energy through water *etc*, wind energy and biofuels are in high demand [2]. But the major drawback with these sources is that they are not reliable because of their intermittent nature. To overcome this drawback, an energy storage device comes as the most reliable method, where the energy stored is used supplied when required. These devices ensure adequate supply of energy timely, and hence are very much reliable. The amount of energy supply depends on the size of energy storage device; large energy storage devices provide energy for many hours even in remote areas whereas small energystorage devices are portable but can supply

---

[*] **Corresponding author Rupesh Rohan:** Indian Rubber Manufacturers Research Association (IRMRA), Sri City Trade Centre, Sri City, District: Chittoor, Andhra Pradesh, India; E-mail: rupeshrohan21@gmail.com

Subhendu Bhandari, Prashant Gupta and Ayan Dey (Eds.)

energy only for limited time period. Large energy storage devices can supply high amount of energy in a short period of time to several places such as where defence installation is done [3].

Natural and synthetic polymers possess a wide range of properties due to which they become the backbone of our daily life. Polymer-based batteries that are used to store energy have gained popularity since decades. These are advantageous as compared to metal originated batteries because of inherent properties of polymer such as light weight and flexibility, high breakdown strength, easy casting of electrodes made up of polymeric materials, vapor deposition, *etc.* These properties make polymers suitable for application in flexible and thinner devices. The electrochemical reaction involved in polymer-based batteries is simple; polymeric electrodes have a lower redox potential but energy density is high. Economically, a polymer can be synthesized at low cost and can also be extracted from bio-based materials. Polymers are used in both energy harvesting and energy storage devices [4].

Along with these advantages, many challenges are associated with polymer-based batteries. Polymers are soluble in electrodes which lead to a threat to battery stability. The active member which travels between the electrodes gets dissolved, leading to reduced cyclability and self-discharge of the batteries [5]. Most of the polymers are of insulating nature, so they require conductive additives which reduce the batteries capacity overall [5].

To overcome the challenges exhibited by polymer batteries, and for extension of their use in energy storage devices, many researches were carried out and polymer composites were developed for that purpose.

Energy storage systems are of two types, electrochemical energy storage devices and non-electrochemical energy storage devices [2]. Fuel cells, batteries, supercapacitors are the types of electrochemical cells and they consist of electrolytes, electrodes and the current collector [6-14]. These components are made of conducting polymers, carbon-based nanomaterials, conducting materials and metal oxides. These materials solve the purpose but are not highly reliable, there are still many challenges with the commercialization of energy storage devices. Since the last decade, research is on-going on developing polymer nanocomposites as an ideal material for making these components. Polymers have high flexibility in designing, because of which they can be used to design different components of an energy storage device [15].

Nanocomposite materials are hybrid materials made up of two or more materials with extremely different chemical and physical properties that remain distinct and separate with a dimension less than 100 nm size range. Nanocomposite materials have two components. The matrix or bulk substance is of one type, while the inorganic nanofiller is the other [15]. The material is called a polymer nanocomposite when a polymer functions as the matrix, and an inorganic nanomaterial works as the nanofiller. Due to the synergic interaction between the polymer and the nanomaterial filler owing to their "nanoeffect," polymer nanocomposite materials have remarkable qualities when compared to polymer composites with micron-size fillers [16-27]. With only a small amount of nanomaterial as a filler, tremendous improvement in the properties of polymeric materials can be achieved. The properties and nature of the polymer employed as the matrix for polymeric nanocomposites (PNCs), as well as the nanofiller, have a significant impact on electric conductivity, processing ease, ionic conductivity, tensile strength, and chemical, thermal, and mechanical stability [16, 23, 26-34].

Polymer nanocomposites have unique physicochemical features that cannot be achieved by using separate components. Due to their intriguing potential for a variety of applications ranging from environmental to medical, polymer nanocomposites have sparked a lot of scientific attention. Sensing and actuation, clean-up, energy storage, electromagnetic (EM) absorption, transportation and safety, defence systems, information technology, and innovative catalysts, among other things. Polymer nanocomposites, in particular, have sparked a lot of attention as a potential solution to both of these problems [19, 24, 27, 31, 32, 35-37].

For the electrochemical application, two types of polymers are generally used:

 i. Electric conducting polymers
ii. Ion conducting polymers

General tendency of a polymer is insulation, but for electrochemical application, polymers having electrical conductivity are in demand. Polyaniline (PANI), polythiopene (PTh), and polypyrrole (PPy) are the polymers having high value in electrochemical systems because they are composed of organic monomers having conjugating double bonds [38, 39]. Along with electrical conductivity, these polymers are also budget friendly, easily processable, light weight, and exhibit thermal as well as mechanical flexibility. Fillers are used in nano form to further enhance the properties of these polymeric materials, and they are called as the nano fillers. Metals, graphene, carbon nanotubes (CNT), carbon, ferromagnetic materials, layered silicates, titanium nanotubes *etc.* are used as the fillers [40]. Polymer nanocomposites have unique physicochemical features that can't be achieved by using single component. Polymer nanocomposites have piqued researchers' interest due to their potential for a wide range of applications, including environmental sustainability, sensing, EM absorption, energy storage and actuation, defence system, safety and transportation. The polymer matrix plays a key role in influencing the processability, transverse tensile properties, compressive properties, shear properties, heat resistance, and resistance to environmental media of PNCs because it is a continuous phase that bonds the nanofillers together [40].

For designing energy storage systems, electrical conductivity of polymers is increased by doping, and chemical or electrochemical redox reactions. In this way, the magnitude of the electrical conductivity is increased several folds.

APPLICATIONS

Lithium-Ion Batteries (LIBs)

Lithium-ion batteries (LIBs) are a key application for PNCs. LIBs have become the most essential and extensively used rechargeable batteries due to their benefits of high working voltage, low toxicity, high capacity, and long cycling life. Silicone, graphite and metal oxides are commonly utilised as anode materials in LIBs; whereas, $LiCoO_2$, $LiFePO_4$, $Li[NiCoAl]O_2$, $Li[MnNiCo]O_2$, and $LiMn_2O_4$ are commonly used as cathode materials. Lithium-ion batteries are widely used to power mobile phones, laptop computers, tablets, and other portable electronic devices. Single lithium-ion cells are connected in series for proper voltage or in parallel to boost the output current in a lithium-ion battery [41-48]. A basic Li-ion cell is made up of a cathode (positive electrode) and anode (negative electrode) separated by an electrolyte and a separator. The separator is made up of a microporous polymer electrolyte that only enables Li ions to pass through, but not the electrons. Carbon, silicon, and metal oxides are the common anode materials, while Li salts like $LiCoO_2$, $LiFePO_4$, and $LiMn_2O_4$ are the common cathode materials. Lithium salts such as $LiClO_4$, $LiPF_6$, $LiAsF_6$, $LiCF_3SO_3$, and $LiBF_4$ in nonaqueous organic solvents such as ethylene carbonate, polyethylene carbonate, or dimethyl carbonate are the usual electrolytes [49, 50].

The electrodes are connected to an external electric supply during the charging process. The electrons are forced to pass through the external circuit from the cathode to the anode. An electrochemical reaction occurs, in which Li^+ ions are produced and transmitted internally to the anode through the electrolyte. The reversible reaction continues until the Li^+ ions are fully intercalated in the anode (LiC_6). During the discharging process, electrons are discharged spontaneously from the anode to the cathode, and Li^+ ions are transported internally *via* the electrolyte from the anode to the cathode. This process continues until all of the LiC_6 is transformed to carbon. Electrolyte is an essential component in Li-ion batteries. Between the electrodes, cathode, and anode lies a layer of electrolyte. Because the electrolyte comes into contact with the electrodes and separator, it interacts electronically with all of the battery's components. The electrolyte aids in the movement of Li^+ ions from the cathode to the anode during charging, as well as from the anode to the cathode during discharge. The electrolyte determines the overall performance of the batteries, including energy density, working potential, working temperature, stability, and cycle life. In order to prevent short circuit in electrodes, it must have a low electronic conductivity. A separator is used to separate the electrodes from the electrolyte. This indicates that the electrolyte has a high ionic conductivity and a low electrical conductivity.

Electrolytes for Lithium-Ion Batteries

The electrolyte is an essential component in Li-ion batteries. The electrolyte is placed between two plates. The electrodes, the cathode, and the electrode, the electrolyte comes into contact with the electricity. It interacts electrically with all of the components in the system, including the switches and the separator. The electrolyte helps the Li^+ ions move from the cathode to the anode during the charging process, and during the discharging process from the anode to the cathode. In general, energy density, working potential, operating temperature, and stability are all factors to consider. The electrolyte determines the battery's durability and cycle life.

So that the electrodes don't short circuit, they have a low electronic conductivity. A separator is used to separate the electrodes from the electrolyte. This indicates that the electrolyte has a high ionic conductivity while having a poor electronic conductivity. For Li-ion batteries, an ideal electrolyte should have:

1. A high ionic conductivity,
2. Large Li^+ ion transference number (close to unity) by reducing the transportation of anions,
3. Excellent mechanical strength,
4. Electrochemical stability over a wide voltage range,
5. Excellent heat and chemical resistance,
6. The electrolyte should be environmentally friendly in the batteries' environment.

Because the electrolyte is in touch with the electrodes, it should have a high conductivity [1, 5, 51]. As a result, only a limited number of materials can be employed as an electrolyte due to chemical deterioration. Liquid electrolytes are currently the most extensively utilised electrolytes. In the case of commercial Li^+ batteries lithium salts are dissolved in dielectric fluids with a high dielectric constant. To achieve a high concentration of Li^+ in the electrolyte, organic solvents were used [1, 5, 51].

To prevent electron transport through the solvent, it must have a high dielectric constant.

To help with the high ionic conductivity, the viscosity should be low. Lithium salts such as $LiClO_4$, $LiPF_6$, $LiAsF_6$, $LiCF_3SO_3$, and $LiBF_4$ make up the majority of the liquid electrolyte which gets dissolved in organic solvents (ethyl acetate, polypropylene acetate, diethyl acetate) tetrahydrofuran, dimethyl carbonate, diethoxyethane, dioxolane, and carbonate). Simple ions such as Cl^-, F^-, or Br^- are not chosen due to their intrinsic properties. Nanocomposites based on polymers are used for energy and environmental applications.

There are two types of liquid electrolyte systems: (1) nonaqueous electrolyte and (2) aqueous electrolyte. The state-of-the-art electrolytes in current commercial Li batteries are nonaqueous electrolyte systems in which lithium salt is dissolved in an organic cyclic carbonate solvent or a mixture [52-64]. In recent decades, there have been ongoing efforts to increase the stability of the carbonate solvent by experimenting with different solvents or salts, or by altering the carbonate salt with additions to improve performance. The synergic interaction of two solvents, which enhances each other's properties, is one technique to boost the electrolyte's performance [49, 50, 65-83].

To deal with all the problems, particularly to achieve high transference number of cation (Li^+), the concept of single ion conducting electrolytes was introduced and as the result of the continuous research conducted, significantly improved performance has been achieved.

Single Ion Conducting Polymer Electrolytes (SIPE) for Lithium-Ion Batteries

The proliferation of portable electronic devices has resulted in a rise in secondary power source demand that is exceedingly difficult to supply with current lithium ion battery technologies. Importance can be given to lithium sulfur batteries and lithium air batteries; they can prove to be the promising innovations for the increasing demand. Liquid electrolytes have high performance in lithium ion batteries, including lithium salts, such as LiBF4, LiPF6, LiAsF6, $LiClO_4$, and lithium bis-(trifluoromethanesulphonyl)imide, dissolved in high dielectric organic solvents like propylene carbonate (PC), ethylene carbonate (EC), and dimethoxyethane (DME). Along with high conductivity, good electrochemical, thermal and mechanical stability is also necessary for optimum performance of batteries. The battery's good thermal stability ensures that it operates within a safe temperature range; the wide electrochemical window prevents side reactions between the electrodes and the electrolyte; and the battery's strong mechanical strength makes battery manufacturing and operation easier [50, 84-93].

However, there are several limitations of lithium based batteries such as lithium salts are not thermally stable in most of the liquid electrolytes, prone to breakdown at high temperatures, and the batteries possess poor electrochemical stability. When compared to dual-ion based electrolytes of polymer-lithium salt systems, single ion conducting polymer electrolytes (SIPEs) are highly unique among all-solid electrolytes discovered to date, with high ionic transference numbers close to unity which minimise the concentration gradient during battery operation to a negligible value [50, 94-98].

Polymer Designing for SIPE

The lithium ion transference number in commercial liquid electrolytes is less than unity because both cation and anion contribute to charge transport in batteries. Reduction in the mobility of big size anions like $N(SO_2CF_3)_2-$(TFSI–) and bis (oxalate) borate (BOB–)raises the lithium ion transference number. However, because the anions are still mobile, the lithium ion transference numbers of these dual ion-based salts, dissolved in organic solvents, are still less than 0.5. Attaching charge delocalized anions to polymer backbones allows anions to be immobilised. The design strategy is shown in Fig (**1**). In SIPE materials, attempts have been made to synthesize charge delocalized sulfonate, bis(sulfonyl)imide, and BOB– (sp^3 boron) anions [99-101]; for example, the chemical structure of bis(sulfonyl) imide based SIPE is shown in Fig (**2**).

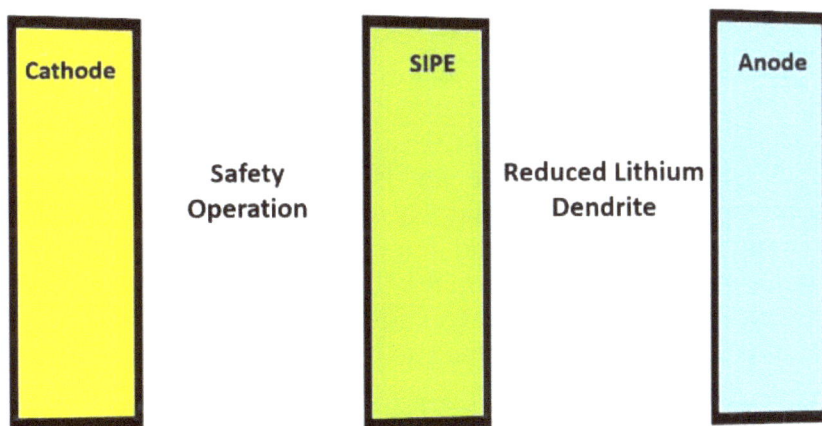

Fig. (1). Schematic diagram for SIPE Li batteries with majorly used functional groups for SIPE based batteries.

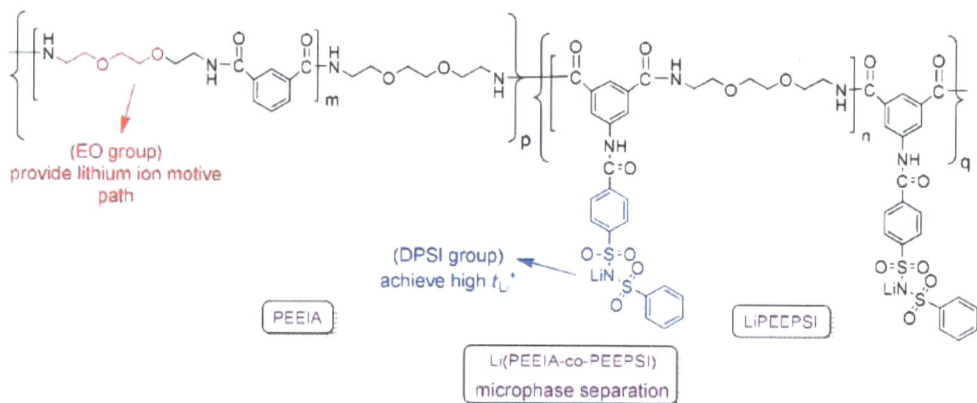

Fig. (2). Chemical Structure one of the SIPE Li batteries with bis(sulfonyl)imide groups [100].

Sulfonate Group Functionalized SIPEs

Nafion, which is the most frequently used commercial single ion polymer in the field of fuel cells, was used as an electrolyte in lithium ion batteries after ion exchange to increase the cyclability. The ionic conductivity of the polymer matrix after the infusion of propylene carbonate (PC) was 4.6310^{-4} S cm^{-1} at room temperature, with good electrochemical stability and mechanical robustness. At an elevated temperature (*i.e.*, 80°C), the performance of the gel Nafion membrane was demonstrated in a half-battery cell with LiFePO$_4$ as a cathode with better discharge cyclability than the battery with LiPF$_6$-EC-DMC (EC: ethylene carbonate, DMC: dimethyl carbonate), which is not surprising given LiPF$_6$'s poor thermal stability compared to Nafion.

In 1999, a sulfonate group-functionalized PEO-based (PEO: polyethylene oxide) SIPE (Poly (ethylene oxide methoxy acrylate-co-lithium 1,1,2 -trifluorobutane sulfonate acrylate) was created. PEO-salt systems have been extensively researched as gel electrolytes. For the trifluorobutane sulfonate single ion conducting polymers, the effect of the oxygen to cation ratio ([O]/[Li$^+$]) on the glass transition temperature (T$_g$) and ionic conductivity was examined using [O]/[Li$^+$] ratios ranging from 12:1 to 28:1. In the PEO-salt systems, there was a rising tendency for the development of ion triplets or higher aggregates that constrain ion mobility to counteract the rise in ionic conductivity as more lithium salt was added. In general, as the [O]/[Li$^+$] ratio in the polymer falls, the supply of charge carriers increases, resulting in an increase in ionic conductivity. When compared to PEO-slat gel complexes, the dependency of ionic conductivity on the [O]/[Li$^+$] ratio is substantially less pronounced for SIPEs. In single-ion polymer systems, this behavior is explained by the creation of ion triplets, and ion aggregates are much decreased. Returning to ionic conductivity, the trifluorobutane sulfonate polymer electrolyte with a 15:1 [O]/[Li$^+$] ratio has an ionic conductivity of 10^{-4} S cm^{-1} at 25°C and 10^{-3} S cm^{-1} at higher temperatures, which is comparable to the conductivity of liquid electrolyte [102-166].

In general, sulfonate group functionalized SIPEs have improved ion conduction after 20 years of research on both molecular structure design and mechanistic study. At room temperature, the conductivity of Li$^+$ has been raised to as high as 10^{-4} S cm^{-1}, which is just one order of magnitude lower than that of the liquid electrolytes. However, instead of commercial Nafion, a battery prototype has not been made employing sulfonate group-functionalized SIPE membranes. Other anion functional groups have received a lot of study attention recently, with bis(sulfonyl)imide and sp^3 boron appearing to be the most successful and widely explored.

Bis(Sulfonyl)imide Group Functionalized SIPEs

In tiny lithium salts for liquid electrolytes, the bis(sulfonyl)imide group has been widely employed. The inductive effect resulting from the electron-withdrawing capability of the two O=S=O groups linked to the nitrogen atom and the trifluoromethyl group is used to achieve high ionic conductivity, as seen in LiN(SO$_2$C$_2$F$_5$)$_2$ (LiBETI) and LiN(SO$_2$CF$_3$)$_2$ (LiTFSI). As a result of the early investigations on ionomers for fuel cell applications, bis(sulfonyl)imide group functionalized SIPE membranes have been widely studied since the 1990s. Bis(sulfonyl)imide was used as a functional group in SIPEs for lithium battery applications in the early 2000s. In a polymer alloy prepared from poly(ethylene oxide-co-propylene oxide) tri acrylate (TA, number average M$_w$=8000) and poly(ethylene oxide-co-propylene oxide) monoacrylate (MA, number average M$_w$=1800), the synthesised poly(2-oxo-1-difuluoroethylene sulfonyl imide) (LiPI) exhibited ionic conductivities ranging from 10^{-7} S cm When compared to monomeric lithium salts in EC at 40°C, the conductivity of the LiPI alloy electrolyte was increased by around 3 orders of magnitude, or 6*10^{-3} S cm^{-1}. With a lithium ion transference number of > 0.7, the poly(5-oxo-3-o-y-4-trifluoromethyl-1,2,4-pentafluoro pentylene sulfonylimide lithium) (LiPPI) membrane exhibits an increased ionic conductivity of 10^{-5} S cm^{-1} at 30°C.

The better ionic conduction performance of LiPI and LiPPI stimulates more research into monomeric lithium salt-functionalized SIPEs for lithium battery applications. Lithium 3- glycidyloxy propanesulfonyl- trifluoro methane sulfonylimide, an ionic epoxy monomer, is used to make networked copolymers (LiGPSI). In both the solid state and the PC swollen state, it has a high ionic conductivity [50, 94-101, 167-237].

sp³ Boron-based SIPEs

Aside from bis(sulfonyl)imide-based SIPEs, sp^3 boron-based SIPEs, are another class of materials with a lot of promise for practical use in lithium batteries. Four oxygen atoms are covalently bound to each boron atom in the sp^3 boron-based SIPEs. Normally, oxygen atoms are linked to electron withdrawing groups in polymer backbones to further promote negative charge delocalization, which is favourable to ion conduction in the polymer matrix. SIPE membranes based on sp^3 boron have excellent ionic conductivity, thermal, and electrochemical stability. The chemical structures of several typical sp^3 boron-based SIPEs were synthesized.

The structure of lithium bis(oxalate) borate (LiBOB), which performs admirably as a liquid electrolyte in lithium ion batteries, was used to create a novel single ion conductor lithium oxalate polyacrylic acid borate for SIPE structural design (LiOPAAB). With an electrochemical window of up to 7.0 V vs. Li$^+$ /Li, better electrochemical stability was achieved. The LiOPAAB membrane has an ambient ionic conductivity of $2.3*10^{-6}$ Scm^{-1} when 3 percent PC solvent was adsorbed. The ambient ionic conductivity was enhanced to $6.11*10^{-6}$ Scm^{-1} by modifying the polymer structure to lithium polyvinyl alcohol oxalate borate (LiPVAOB) and raising the PC content to 20% due to the superior electrochemical stability [238-277].

Improvements in ionic conductivity, thermal, and electrochemical stability in SIPE materials have been made during the last 20 years to fulfill the increasingly severe requirements for realistic battery operations. Although most of the prototypes are fitted with gel SIPEs, it has been proved that SIPEs can withstand far greater temperatures than monomeric lithium salts in liquid electrolytes during battery operation. Nonetheless, the challenges of operating SIPE-based lithium batteries at room temperature and in an all-solid state still remain there and are to be resolved. However, Sp^3 boron and delocalized sulfonate group-based SIPE with a flexible polymer framework could be possible avenues. An example of Sp^3 boron based SIPE is shown in Fig (**3**) [5, 214, 278-285].

Supercapacitors

The future of global energy will be determined by the consumers and the industries utilizing and creating energy more effectively. Supercapacitors have already shown to be one of the leading technologies for intermittent storage and high-power delivery. Supercapacitors are one of the intermittents which are being used to store charges. Analogous to lithium-ion batteries, the supercapacitors are known for their high energy and power density as demonstrated in Ragone plot, for their promising potentials for future systems including portable electronics, hybrid electric vehicles, and large industrial equipment. The electrochemical capacitors are expected to fill in the gap between the batteries and the conventional electrostatic capacitors, where the former has high energy density but relatively low power density; whereas, the latter has high power density but too low energy density. As an energy storage device, supercapacitors have ultrahigh capacitance, high power density, and long cycle. Supercapacitor materials frequently require high specific surface area, mechanical and chemical resilience, and low cost. Supercapacitors are used as supplemental energy sources in consumer electronics, electric automobiles and hybrid electric vehicles, electric utilities, and industrial areas due to their quick charge/discharge and high power density. However, because of their size, it is challenging for supercapacitors to be used as a major power source. Low energy densities and fast self-discharge innovations are being used to address these concerns. Essentials for supercapacitor designs are electrode materials and electrolytes. Furthermore, developing efficient battery-supercapacitor hybrids would be appealing. Hybrid systems combine the benefits of quick charge/discharge and energy storage. Supercapacitors having a high power density as well as a high energy density exhibit low self-discharge [286-288].

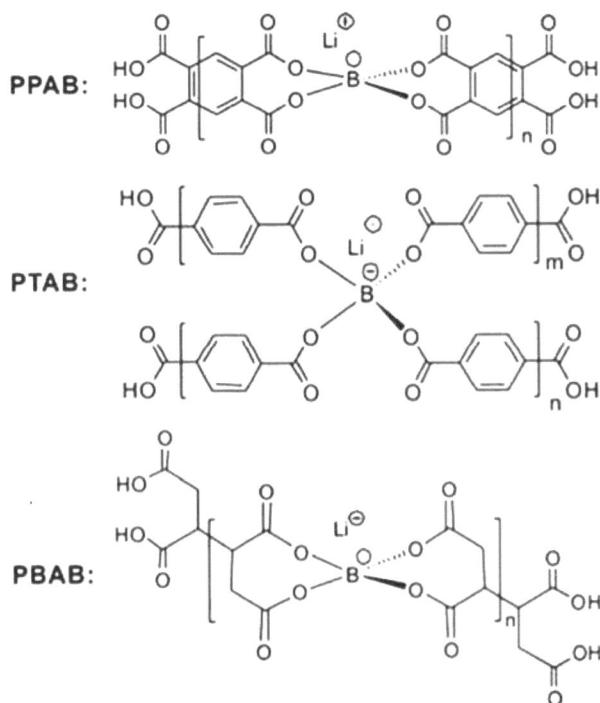

Fig. (3). Examples of sp^3 boron-based polymeric compounds for SIPE [282].

Electric double-layer capacitors (EDLCs), in which charges are stored by the formation of an electric double-layer (EDL) at the electrode/electrolyte interface, and pseudocapacitors (PCs), which have pseudocapacitance due to oxidation–reduction reactions, are the two types of electrochemical supercapacitors. Furthermore, hybrid capacitors were created by combining the two types of charge storage mechanisms. Electrode materials, particularly conductivity and accessible surface areas, have a big impact on their performance.

SOHIO and NEC were the first to commercialize EDLCs, also known as electrostatic supercapacitors with symmetrical electrodes as cathode and anode. Furthermore, substantial R&D efforts to improve EDLCs have been conducted [47-49]. Traditional EDLCs use activated carbon as an electrode material, which has a capacitance of 40–70 F/g. Pseudocapacitors are also known as the faradaic supercapacitors in computers. Electrodes are made of transition metal oxides and conducting polymers. Low electrical conductivity, insufficient use of transition metal oxides, and poor cycling stability of conducting polymers, on the other hand, result in adverse rate conditions performance and density of power. Unlike EDLCs and PCs, which have symmetrical structures, hybrid capacitors have an asymmetrical layout in which an electric double layer carbon material is utilised as an anode and a pseudocapacitive material is employed as a cathode. The several types of supercapacitors and their applications in consumer electronics, automobiles, electric utilities, and industry will be reviewed.

Components of Supercapacitors

Current Collectors

A current collector is an essential component of supercapacitors since it not only supports electrode materials but also links electrodes to the capacitor terminal. The capacitance, energy density, power density, and cycle life of supercapacitors are all influenced by the contact impedance between electrodes and current collectors [288, 289].

Electrolytes have a big role in choosing current collectors. The most common current collector is titanium (or Au) foil. The other current collectors include nickel foil (or nickel foam) as working electrode in acidic as well as in alkaline aqueous electrolytes, and aluminium foil in nonaqueous electrolytes. Ideal nonaqueous solution current collectors should have good interface conductivity and electrochemical stability at high potentials. Current collectors' surfaces can be treated to reduce ohmic drops at the contact, and coatings on them can improve their electrochemical stability at high potential. Current collectors with nanostructured structures are being developed. Increase of the contact area between active materials and current collectors improve the interaction. In addition, the expanding demand for wearable electronic devices and flexible supercapacitors has fuelled the development of flexible supercapacitor collectors in the present. Carbon in the form of a highly conductive nanotube or graphene are the two examples. Metallic textiles, polymer film, and Ni foam are the examples of these materials [290-306].

Separators

Glass fibres, cellulose, and polymer membranes are the common separators because they are thin and extremely porous membranes or films. In a supercapacitor cell, the membranes serve to separate the cathode and anode. Separators that qualify must have low ion transfer resistance, excellent electronic insulating properties, strong electrochemical stability and excellent mechanical toughness. The types of electrodes, working temperature, and other factors influence separator selection as well as the possibility for operation. In addition, the shape, pore structure, thickness, and chemical content of the surface. Separators have a significant impact on particular capacitance, specific energy and power densities, electronic series resistance, and polarizability limitations. New separator materials have been investigated in addition to existing separator materials. GO films and eggshell membranes are the examples of supercapacitors [307].

Binders

Fluorinated polymetric materials like poly(vinylidenefluoride) (PVDF) and polytetrafluoroethylene (PTFE) are commonly employed as the binders in supercapacitor cells to improve the electrical contact of the active materials and their adhesion to the current collectors. Binders can increase the mechanical and electrical properties of electrodes, albeit at the expense of ion diffusion. As a result, it is critical to get the binder content in electrodes just right. Furthermore, supercapacitors' performance can be substantially improved. Binder characteristics have an impact on supercapacitor performance. The use of PTFE as a binder in a negative electrode layer improved the results. PVDF has a better energy and power performance. The use of polycarboxylate instead of PVDF as a binder can reduce the initially observed effects. Potassium-ion capacitors have an irreversible capacity [288, 308, 309].

Electrolytes

The energy density and power density of supercapacitors are both affected by electrolytes. The following characteristics should be present in an ideal electrolyte for electrochemical supercapacitor (ES): (a) a wide voltage window, (b) high electrochemical stability, (c) large ionic concentration, (d) small solvated ionic radius, (e) low resistance, and (f) others (such as low viscosity, volatility, toxicity, low-temperature property, and cost). Aqueous electrolytes, organic electrolytes, and ionic liquids are the three forms of electrolytes for ES cells [310-328].

Though there are different types of materials, which have completely different material and performance properties, covering inorganic to organic, put together to make supercapacitors as well lithium ion batteries, the role of polymers has become very prominent as well as a compulsion in such devices. From binders to solid electrolytes (which are future of energy storage system), polymers have been considered significant. For commercialization, where cost becomes a deciding factor, supercapacitors and batteries cannot be fabricated without polymers [329, 330].

CONCLUSION

In this chapter, we have attempted to outline the most recent developments in polymer nanocomposite-based material development for electrical energy storage systems (EESSs) like secondary lithium-ion batteries and supercapacitors. Electrodes made of polymer nanocomposites and the electrolyte in Li-ion batteries, as well as the electrode in supercapacitors, are essential to do this. All-plastic, flexible, wearable electric energy storage devices are the stuff of dreams. A significant amount of research has gone into developing all-solid, flexible energy, gadgets for storing portable devices. The use of cutting-edge nanomaterials such as carbon nanotubes, graphene, metal organic frameworks, and mesoporous carbon has greatly aided the creation of materials. However, these materials raise challenges such as commercial production, environmental concerns, and pricing. A polymer backbone matrix and an inorganic filler, such as nanomaterials, make up a polymer nanocomposite. Because of the synergistic interaction between the components, they have great qualities. The polymers in these polymer nanocomposites have strong electric conductivity and flexibility, while the fillers have good stability and dielectric characteristics. However, optimising desired qualities in a polymer nanocomposite still necessitates extensive theoretical and experimental study to establish structure-property relationships and create materials with unique properties. Furthermore, more efforts in synthetic chemistry are needed to create polymers and polymer nanocomposite with a controlled structure. As a result, the development of high energy density and high power output materials for supercapacitors and lithium-ion batteries with long-term cycle stability is critical in realizing the dream of replacing fossil fuels for the betterment of the environment and to ensure energy for the future.

ACKNOWLEDGEMENT

Declared none.

REFERENCES

[1]   S. K. S. Hossain and M. E. Hoque, 9 - Polymer nanocomposite materials in energy storage: Properties and applications, in Polymer-based Nanocomposites for Energy and Environmental Applications, M. Jawaid and M. M. Khan, Eds., ed: Woodhead Publishing, 2018, pp. 239-282.
        [http://dx.doi.org/10.1016/B978-0-08-102262-7.00009-X]

[2]   A. A. K. Arani, H. Karami, G. B. Gharehpetian, and M. S. A. Hejazi, Review of flywheel energy storage systems structures and applications in power systems and microgrids, Renewable and Sustainable Energy Reviews, 69, pp. 9-18, 2017.
        [http://dx.doi.org/10.1016/j.rser.2016.11.166]

[3]   B. C. Riggs, S. Adireddy, C. H. Rehm, V. S. Puli, R. Elupula, and D. B. Chrisey, Polymer nanocomposites for energy storage applications, materials today: Proceedings, 2, pp. 3853-3863, 2015.
        [http://dx.doi.org/10.1016/j.matpr.2015.08.004]

[4]   Yang C, Wei H, Guan L, *et al.* Polymer nanocomposites for energy storage, energy saving, and anticorrosion. J Mater Chem A Mater Energy Sustain 2015; 3(29): 14929-41.
        [http://dx.doi.org/10.1039/C5TA02707A]

[5]   Rohan R, Sun Y, Cai W, *et al.* Functionalized meso/macro-porous single ion polymeric electrolyte for applications in lithium ion batteries. J Mater Chem A Mater Energy Sustain 2014; 2(9): 2960-7.
        [http://dx.doi.org/10.1039/C3TA13765A]

[6]   Arbizzani C, Mastragostino M, Meneghello L. Polymer-based redox supercapacitors: A comparative study. Electrochim Acta 1996; 41(1): 21-6.
        [http://dx.doi.org/10.1016/0013-4686(95)00289-Q]

[7]   Asensio JA, Borrós S, Gómez-Romero P. Enhanced conductivity in polyanion-containing polybenzimidazoles. Improved materials for proton-exchange membranes and PEM fuel cells. Electrochem Commun 2003; 5(11): 967-72.
        [http://dx.doi.org/10.1016/j.elecom.2003.09.007]

[8]   M. Ates and I. Ekmen, Polym.-Plast. Technol. Engineer., 2016, 55, 2016. p. 1489.

[9]   Baril D, Michot C, Armand M. Electrochemistry of liquids vs. solids: Polymer electrolytes. Solid State Ion 1997; 94(1-4): 35-47.
        [http://dx.doi.org/10.1016/S0167-2738(96)00614-5]

[10]   Barisci JN, Conn C, Wallace GG. Trend Polym Sci 1996; 9: 307.

[11]   Bartlett PN, Whitaker RG. Electrochemical immobilisation of enzymes. J Electroanal Chem Interfacial Electrochem 1987; 224(1-2): 27-35.
        [http://dx.doi.org/10.1016/0022-0728(87)85081-7]

[12]   Bélanger D, Ren X, Davey J, Uribe F, Gottesfeld S. Characterization and long-term performance of polyaniline-based electrochemical capacitors. J Electrochem Soc 2000; 147(8): 2923.
        [http://dx.doi.org/10.1149/1.1393626]

[13]   Bockris JOM, Kita H. Analysis of galvanostatic transients and application to the iron electrode reaction. J Electrochem Soc 1961; 108(7): 676.
        [http://dx.doi.org/10.1149/1.2428188]

[14]   Cindrella L, Kannan AM. Membrane electrode assembly with doped polyaniline interlayer for proton exchange membrane fuel cells under low relative humidity conditions. J Power Sources 2009; 193(2): 447-53.
        [http://dx.doi.org/10.1016/j.jpowsour.2009.04.002]

[15]   D. R. Paul and L. M. Robeson, Polymer nanotechnology: nanocomposites, polymer, 49, pp. 3187-3204, 2008.
       [http://dx.doi.org/10.1016/j.polymer.2008.04.017]

[16]   A. J. Crosby and J. Y. Lee, Polymer nanocomposites: the nano effect on mechanical properties, polymer reviews, 47, pp. 217-229, 2007.

[17]   Donald AM, Kramer EJ. Effect of molecular entanglements on craze microstructure in glassy polymers. J Polym Sci, Polym Phys Ed 1982; 20(5): 899-909.
       [http://dx.doi.org/10.1002/pol.1982.180200512]

[18]   Fornes TD, Yoon PJ, Paul DR. Polymer matrix degradation and color formation in melt processed nylon 6/clay nanocomposites. Polymer (Guildf) 2003; 44(24): 7545-56.
       [http://dx.doi.org/10.1016/j.polymer.2003.09.034]

[19]   Fox TG. Bull Am Phys Soc 1956; 2: 123.

[20]   Gupta S, Zhang Q, Emrick T, Balazs AC, Russell TP. Entropy-driven segregation of nanoparticles to cracks in multilayered composite polymer structures. Nat Mater 2006; 5(3): 229-33.
       [http://dx.doi.org/10.1038/nmat1582]

[21]   Ishida H, Campbell S, Blackwell J. General approach to nanocomposite preparation. Chem Mater 2000; 12(5): 1260-7.
       [http://dx.doi.org/10.1021/cm990479y]

[22]   S. Kawana and R. A. L. Jones, Physical Review E, 6302, p. null, 2001.

[23]   B. D. Lauterwasser and E. J. Kramer, Philosophical Magazine a-Physics of Condensed Matter Structure Defects and Mechanical Properties, 39, p. 469, 1979.

[24]   Lee JY, Buxton GA, Balazs AC. Using nanoparticles to create self-healing composites. J Chem Phys 2004; 121(11): 5531-40.
       [http://dx.doi.org/10.1063/1.1784432] [PMID: 15352848]

[25]   Lee JY, Crosby AJ. Crazing in glassy block copolymer thin films. Macromolecules 2005; 38(23): 9711-7.
       [http://dx.doi.org/10.1021/ma051716n]

[26]   Lee JY, Zhang Q, Emrick T, Crosby AJ. Nanoparticle alignment and repulsion during failure of glassy polymer nanocomposites. Macromolecules 2006; 39(21): 7392-6.
       [http://dx.doi.org/10.1021/ma061210k]

[27]   Lin Y, Böker A, He J, *et al.* Self-directed self-assembly of nanoparticle/copolymer mixtures. Nature 2005; 434(7029): 55-9.
       [http://dx.doi.org/10.1038/nature03310] [PMID: 15744296]

[28]   Hsiao CC, Lin TS, Cheng LY, Ma CCM, Yang ACM. The nanomechanical properties of polystyrene thin films embedded with surface-grafted multiwalled carbon nanotubes. Macromolecules 2005; 38(11): 4811-8.
       [http://dx.doi.org/10.1021/ma048413y]

[29]   Huh J, Ginzburg VV, Balazs AC. Thermodynamic behavior of particle/diblock copolymer mixtures: simulation and theory. Macromolecules 2000; 33(21): 8085-96.
       [http://dx.doi.org/10.1021/ma000708y]

[30]   Liu H, Brinson LC. A hybrid numerical-analytical method for modeling the viscoelastic properties of polymer nanocomposites. J Appl Mech 2006; 73(5): 758-68.
       [http://dx.doi.org/10.1115/1.2204961]

[31]   C. M. Stafford, S. Guo, C. Harrison, and M. Y. M. Chiang, Review of Scientific Instruments, 76, p. null, 2005.

[32]   Stafford CM, Harrison C, Beers KL, *et al.* A buckling-based metrology for measuring the elastic

moduli of polymeric thin films. Nat Mater 2004; 3(8): 545-50.
[http://dx.doi.org/10.1038/nmat1175] [PMID: 15247909]

[33]   Varlot K, Reynaud E, Kloppfer MH, Vigier G, Varlet J. Clay-reinforced polyamide: Preferential
        orientation of the montmorillonite sheets and the polyamide crystalline lamellae. J Polym Sci, B,
        Polym Phys 2001; 39(12): 1360-70.
        [http://dx.doi.org/10.1002/polb.1108]

[34]   Zhang Q, Gupta S, Emrick T, Russell TP. Surface-functionalized CdSe nanorods for assembly in
        diblock copolymer templates. J Am Chem Soc 2006; 128(12): 3898-9.
        [http://dx.doi.org/10.1021/ja058615p] [PMID: 16551083]

[35]   Saujanya C, Radhakrishnan S. Structure development and crystallization behaviour of
        PP/nanoparticulate composite. Polymer (Guildf) 2001; 42(16): 6723-31.
        [http://dx.doi.org/10.1016/S0032-3861(01)00140-9]

[36]   Shah D, Maiti P, Jiang DD, Batt CA, Giannelis EP. Effect of nanoparticle mobility on toughness of
        polymer nanocomposites. Adv Mater 2005; 17(5): 525-8.
        [http://dx.doi.org/10.1002/adma.200400984]

[37]   Stafford CM, Vogt BD, Harrison C, Julthongpiput D, Huang R. Elastic moduli of ultrathin amorphous
        polymer films. Macromolecules 2006; 39(15): 5095-9.
        [http://dx.doi.org/10.1021/ma060790i]

[38]   Yang J, Liu Y, Liu S, Li L, Zhang C, Liu T. Conducting polymer composites: material synthesis and
        applications in electrochemical capacitive energy storage. Mater Chem Front 2017; 1(2): 251-68.
        [http://dx.doi.org/10.1039/C6QM00150E]

[39]   Niu Z, Luan P, Shao Q, *et al.* A skeleton/skin strategy for preparing ultrathin free-standing single-
        walled carbon nanotube/polyaniline films for high performance supercapacitor electrodes. Energy
        Environ Sci 2012; 5(9): 8726-33.
        [http://dx.doi.org/10.1039/c2ee22042c]

[40]   G. A. Snook, P. Kao, and A. S. Best, Conducting-polymer-based supercapacitor devices and
        electrodes, Journal of Power Sources, 196, pp. 1-12, 2011.
        [http://dx.doi.org/10.1016/j.jpowsour.2010.06.084]

[41]   Tian C, Lin F, Doeff MM. Electrochemical characteristics of layered transition metal oxide cathode
        materials for lithium ion batteries: surface, bulk behavior, and thermal properties. Acc Chem Res
        2018; 51(1): 89-96.
        [http://dx.doi.org/10.1021/acs.accounts.7b00520] [PMID: 29257667]

[42]   Zhao E, Fang L, Chen M, *et al.* New insight into Li/Ni disorder in layered cathode materials for
        lithium ion batteries: a joint study of neutron diffraction, electrochemical kinetic analysis and first-
        principles calculations. J Mater Chem A Mater Energy Sustain 2017; 5(4): 1679-86.
        [http://dx.doi.org/10.1039/C6TA08448F]

[43]   Chakraborty A, Dixit M, Aurbach D, Major DT. Predicting accurate cathode properties of layered
        oxide materials using the SCAN meta-GGA density functional, npj Comput. Mater 2018; 4: 60.

[44]   Lin F, Nordlund D, Markus IM, Weng TC, Xin HL, Doeff MM. Profiling the nanoscale gradient in
        stoichiometric layered cathode particles for lithium-ion batteries. Energy Environ Sci 2014; 7(9):
        3077.
        [http://dx.doi.org/10.1039/C4EE01400F]

[45]   Xu X, Lee S, Jeong S, Kim Y, Cho J. Recent progress on nanostructured 4V cathode materials for Li-
        ion batteries for mobile electronics. Mater Today 2013; 16(12): 487-95.
        [http://dx.doi.org/10.1016/j.mattod.2013.11.021]

[46]   Chakraborty A, Kunnikuruvan S, Kumar S, *et al.* Layered cathode materials for lithium-ion batteries:
        review of computational studies on $LiNi_{1-x-y}Co_xMn_yO^2$ and $LiNi_{1-x-y}Co_xAl_yO^2$. Chemistry of
        Materials 2020; 32: 915-52.

[47]   Zheng J, Myeong S, Cho W, *et al.* Li- and Mn-Rich cathode materials: challenges to commercialization. Adv Energy Mater 2017; 7(6): 1601284.
[http://dx.doi.org/10.1002/aenm.201601284]

[48]   Whittingham MS. Lithium batteries and cathode materials. Chem Rev 2004; 104(10): 4271-302.
[http://dx.doi.org/10.1021/cr020731c] [PMID: 15669156]

[49]   The and R. P. T. Tomkins, Nonaqueous Electrolytes Handbook 1, 1972.

[50]   K. Xu, Nonaqueous Liquid Electrolytes for Lithium-Based Rechargeable Batteries, Chemical Reviews, 104, pp. 4303-4418, 2004.

[51]   Rohan R, Pareek K, Chen Z, *et al.* A high performance polysiloxane-based single ion conducting polymeric electrolyte membrane for application in lithium ion batteries. J Mater Chem A Mater Energy Sustain 2015; 3(40): 20267-76.
[http://dx.doi.org/10.1039/C5TA02628H]

[52]   J. B. Goodenough and Y. Kim, Challenges for rechargeable li batteries, chemistry of materials, 22, pp. 587-603, 2010.

[53]   J. B. Goodenough and Y. J. Kim, Solid State Chem.

[54]   J. B. Goodenough, K. Mizushima, and T. Takeda, Jpn. J. Appl. Phys., 19−3, p. 305, 1980.

[55]   Goodenough J B, van Schalkwijk W, Scrosati B. Advances in Li-Ion Batteries 2002.

[56]   Guerfi A, Dontigny M, Kobayashi Y, Vijh A, Zaghib K. Investigations on some electrochemical aspects of lithium-ion ionic liquid/gel polymer battery systems. J Solid State Electrochem 2009; 13(7): 1003-14.
[http://dx.doi.org/10.1007/s10008-008-0697-x]

[57]   Imhof R, Novák P. Oxidative electrolyte solvent degradation in lithium-ion batteries: an *in situ* differential electrochemical mass spectrometry investigation. J Electrochem Soc 1999; 146(5): 1702-6.
[http://dx.doi.org/10.1149/1.1391829]

[58]   Inda Y, Katoh T, Baba M. Development of all-solid lithium-ion battery using Li-ion conducting glass-ceramics. J Power Sources 2007; 174(2): 741-4.
[http://dx.doi.org/10.1016/j.jpowsour.2007.06.234]

[59]   Kang K, Carlier D, Reed J, *et al.* Synthesis and electrochemical properties of layered $Li_{0.9} Ni_{0.45} Ti_{0.55} O_2$. Chem Mater 2003; 15(23): 4503-7.
[http://dx.doi.org/10.1021/cm034455+]

[60]   Kim JH, Myung ST, Yoon CS, Oh IH, Sun YK. Effect of ti substitution for mn on the structure of $lini_{0.5}mn_{1.5-x}ti_xo_4$ and their electrochemical properties as lithium insertion material. J Electrochem Soc 2004; 151(11): A1911.
[http://dx.doi.org/10.1149/1.1805524]

[61]   Kim Y, Park K, Song S, Han J, Goodenough JB. Access to m^{3+}/m^{2+} redox couples in layered $lims_2$ sulfides (m=ti, v, cr) as anodes for li-ion battery. J Electrochem Soc 2009; 156(8): A703.
[http://dx.doi.org/10.1149/1.3151856]

[62]   Liu J, Manthiram A. Understanding the improvement in the electrochemical properties of surface modified 5 V $LiMn_{1.42} Ni_{0.42} Co_{0.16} o_4$ spinel cathodes in lithium-ion cells. Chem Mater 2009; 21(8): 1695-707.
[http://dx.doi.org/10.1021/cm9000043]

[63]   Lu Z, Dahn JR. Understanding the anomalous capacity of $li/li[ni_xli_{(1/3-2x/3)}mn_{(2/3-x/3)}]o_2$ cells using *in situ* x-ray diffraction and electrochemical studies. J Electrochem Soc 2002; 149(7): A815.
[http://dx.doi.org/10.1149/1.1480014]

[64] Nagata K, Nanno T. All solid battery with phosphate compounds made through sintering process. J Power Sources 2007; 174(2): 832-7.
[http://dx.doi.org/10.1016/j.jpowsour.2007.06.227]

[65] Arora P, White RE, Doyle M. Capacity Fade Mechanisms and Side Reactions in Lithium-Ion Batteries. J Electrochem Soc 1998; 145(10): 3647-67.
[http://dx.doi.org/10.1149/1.1838857]

[66] Bittihn R, Herr R, Hoge D. The SWING system, a nonaqueous rechargeable carbon/metal oxide cell. J Power Sources 1993; 43(1-3): 223-31.
[http://dx.doi.org/10.1016/0378-7753(93)80118-9]

[67] Blomgren GE. Liquid electrolytes for lithium and lithium-ion batteries. J Power Sources 2003; 119-121: 326-9.
[http://dx.doi.org/10.1016/S0378-7753(03)00147-2]

[68] Dey AN. Lithium anode film and organic and inorganic electrolyte batteries. Thin Solid Films 1977; 43(1-2): 131-71.
[http://dx.doi.org/10.1016/0040-6090(77)90383-2]

[69] Ebner W, Fouchard D, Xie L. The $LiNiO^2$/carbon lithium-ion battery. Solid State Ion 1994; 69(3-4): 238-56.
[http://dx.doi.org/10.1016/0167-2738(94)90413-8]

[70] Fong R, von Sacken U, Dahn JR. Studies of lithium intercalation into carbons using nonaqueous electrochemical cells. J Electrochem Soc 1990; 137(7): 2009-13.
[http://dx.doi.org/10.1149/1.2086855]

[71] Foos JS, Stolki TJ. A new ether solvent for lithium cells. J Electrochem Soc 1988; 135(11): 2769-71.
[http://dx.doi.org/10.1149/1.2095427]

[72] Koch VR, Young JH. The stability of the secondary lithium electrode in tetrahydrofuran-based electrolytes. J Electrochem Soc 1978; 125(9): 1371-7.
[http://dx.doi.org/10.1149/1.2131680]

[73] Nagaura T, Ozawa K. Prog Batteries Sol Cells 1990; 9: 209.

[74] Nishi Y, Azuma H, Omaru AUS. Patent 4,959,281, 1990.

[75] Ohzhku T, Ueda A, Nagayama M. J Electrochem Soc 1993; 140: 1862.
[http://dx.doi.org/10.1149/1.2220730]

[76] Rauh RD, Brummer SB. The effect of additives on lithium cycling in methyl acetate. Electrochim Acta 1977; 22(1): 85-91.
[http://dx.doi.org/10.1016/0013-4686(77)85058-5]

[77] Xu K, Angell CA. Synthesis and characterization of lithium sulfonates as components of molten salt electrolytes. Electrochim Acta 1995; 40(13-14): 2401-3.
[http://dx.doi.org/10.1016/0013-4686(95)00203-Q]

[78] Xu K, Ding MS, Zhang S, Allen JL, Jow TR. An attempt to formulate nonflammable lithium ion electrolytes with alkyl phosphates and phosphazenes. J Electrochem Soc 2002; 149(5): A622.
[http://dx.doi.org/10.1149/1.1467946]

[79] Xu K, Zhang S, Allen JL, Jow TR. Evaluation of fluorinated alkyl phosphates as flame retardants in electrolytes for li-ion batteries: ii. performance in cell. J Electrochem Soc 2003; 150(2): A170.
[http://dx.doi.org/10.1149/1.1533041]

[80] Xu W, Angell CA. Weakly coordinating anions, and the exceptional conductivity of their nonaqueous solutions. Electrochem Solid-State Lett 2001; 4(1): E1.
[http://dx.doi.org/10.1149/1.1344281]

[81] Yazami R, Guérard D. Some aspects on the preparation, structure and physical and electrochemical

properties of LixC6. J Power Sources 1993; 43(1-3): 39-46.
[http://dx.doi.org/10.1016/0378-7753(93)80100-4]

[82] Zhang SS, Jow TR. Aluminum corrosion in electrolyte of Li-ion battery. J Power Sources 2002;
109(2): 458-64.
[http://dx.doi.org/10.1016/S0378-7753(02)00110-6]

[83] Zhang SS, Xu K, Jow TR. Understanding formation of solid electrolyte interface film on limn[sub
2]o[sub 4] electrode. J Electrochem Soc 2002; 149(12): A1521.
[http://dx.doi.org/10.1149/1.1516220]

[84] Blomgren GE. Liquid electrolytes for lithium and lithium-ion batteries. J Power Sources 2003; 119-
121: 326-9.
[http://dx.doi.org/10.1016/S0378-7753(03)00147-2]

[85] Aravindan V, Gnanaraj J, Madhavi S, Liu HK. Lithium-ion conducting electrolyte salts for lithium
batteries. Chemistry 2011; 17(51): 14326-46.
[http://dx.doi.org/10.1002/chem.201101486] [PMID: 22114046]

[86] Selim R, Bro P. Some observations on rechargeable lithium electrodes in a propylene carbonate
electrolyte. J Elec Soci 1974; 121: 1457-9.
[http://dx.doi.org/10.1149/1.2401708]

[87] Koch V R, Goldman J L, Mattos C J, Mulvaney M. Specular lithium deposits from lithium
hexafluoroarsenate/diethyl ether electrolytes. J Elec Soc 1982; 129: 1-4.
[http://dx.doi.org/10.1149/1.2123756]

[88] Pistoia G, Rossi M D, Scrosati B. Study of the behavior of ethylene carbonate as a nonaqueous battery
solvent. J Elec Soc 1970; 117: 500-2.
[http://dx.doi.org/10.1149/1.2407550]

[89] Pistoia G. Nonaqueous batteries with liclo4-ethylene carbonate as electrolyte. J Elec Soc 1971; 118:
153-8.

[90] McMillan R S, Juzkow M W. A report on the development of a rechargeable lithium cell for
application in autofocus cameras. J Elec Soc 1991; 138: 1566-9.
[http://dx.doi.org/10.1149/1.2085834]

[91] Tobishima S, Arakawa M, Hirai T, Yamaki J. Ethylene carbonate-based electrolytes for rechargeable
lithium batteries. J Power Sources 1989; 26(3-4): 449-54.
[http://dx.doi.org/10.1016/0378-7753(89)80162-4]

[92] Tarascon J M, Guyomard D. New electrolyte compositions stable over the 0 to 5 V voltage range and
compatible with the Li1+xMn2O4/carbon Li-ion cells. Solid State Ionics 1994; 69: 293-305.
[http://dx.doi.org/10.1016/0167-2738(94)90418-9]

[93] Lazzari M. A cyclable lithium organic electrolyte cell based on two intercalation electrodes. J Elec Soc
1980; 127: 773.
[http://dx.doi.org/10.1149/1.2129753]

[94] D. Golodnitsky. Electrolytes: Single lithium ion conducting polymers. 2009, 5.

[95] Lin KJ, Li K, Maranas JK. Differences between polymer/salt and single ion conductor solid polymer
electrolytes. RSC Advances 2013; 3(5): 1564-71.
[http://dx.doi.org/10.1039/C2RA21644B]

[96] Zhang SS. A review on the separators of liquid electrolyte Li-ion batteries. J Power Sources 2007;
164(1): 351-64.
[http://dx.doi.org/10.1016/j.jpowsour.2006.10.065]

[97] X.-G. Sun and J. B. Kerr, Synthesis and characterization of network single ion conductors based on
comb-branched polyepoxide ethers and lithium bis(allylmalonato)borate, macromolecules, 39, pp.
362-372, 2005.

[98]   Meyer WH. Polymer electrolytes for lithium-ion batteries. Adv Mater 1998; 10(6): 439-48.
[http://dx.doi.org/10.1002/(SICI)1521-4095(199804)10:6<439::AID-ADMA439>3.0.CO;2-I] [PMID: 21647973]

[99]   Xu G, Zhang Y, Rohan R, Cai W, Cheng H. Synthesis, characterization and battery performance of a lithium poly (4-vinylphenol) phenolate borate composite membrane. Electrochim Acta 2014; 139: 264-9.
[http://dx.doi.org/10.1016/j.electacta.2014.06.173]

[100] Zhang Y, Lim CA, Cai W, *et al.* Design and synthesis of a single ion conducting block copolymer electrolyte with multifunctionality for lithium ion batteries. RSC Advances 2014; 4(83): 43857-64.
[http://dx.doi.org/10.1039/C4RA08709G]

[101] Qin B, Liu Z, Zheng J, *et al.* Single-ion dominantly conducting polyborates towards high performance electrolytes in lithium batteries. J Mater Chem A Mater Energy Sustain 2015; 3(15): 7773-9.
[http://dx.doi.org/10.1039/C5TA00216H]

[102] Abraham KM, Jiang Z, Carroll B. Highly conductive peo-like polymer electrolytes. Chem Mater 1997; 9(9): 1978-88.
[http://dx.doi.org/10.1021/cm970075a]

[103] Alfaruqi MH, Gim J, Kim S, *et al.* A layered δ-MnO$_2$ nanoflake cathode with high zinc-storage capacities for eco-friendly battery applications. Electrochem Commun 2015; 60: 121-5.
[http://dx.doi.org/10.1016/j.elecom.2015.08.019]

[104] Alfaruqi MH, Islam S, Putro DY, *et al.* Structural transformation and electrochemical study of layered MnO^2 in rechargeable aqueous zinc-ion battery. Electrochim Acta 2018; 276: 1-11.
[http://dx.doi.org/10.1016/j.electacta.2018.04.139]

[105] Alfaruqi MH, Mathew V, Gim J, *et al.* Electrochemically induced structural transformation in a γ-mno$_2$ cathode of a high capacity zinc-ion battery system. Chem Mater 2015; 27(10): 3609-20.
[http://dx.doi.org/10.1021/cm504717p]

[106] Anothumakkool B, Kurungot S. Electrochemically grown nanoporous MnO^2 nanowalls on a porous carbon substrate with enhanced capacitance through faster ionic and electrical mobility. Chem Commun (Camb) 2014; 50(54): 7188-90.
[http://dx.doi.org/10.1039/c4cc00927d] [PMID: 24865591]

[107] Aziz SB, Woo TJ, Kadir MFZ, Ahmed HM. A conceptual review on polymer electrolytes and ion transport models. J Sci Adv Mater Devices 2018; 3(1): 1-17.
[http://dx.doi.org/10.1016/j.jsamd.2018.01.002]

[108] Balakrishnan PG, Ramesh R, Prem Kumar T. Safety mechanisms in lithium-ion batteries. J Power Sources 2006; 155(2): 401-14.
[http://dx.doi.org/10.1016/j.jpowsour.2005.12.002]

[109] Bieker G, Winter M, Bieker P. Electrochemical *in situ* investigations of SEI and dendrite formation on the lithium metal anode. Phys Chem Chem Phys 2015; 17(14): 8670-9.
[http://dx.doi.org/10.1039/C4CP05865H] [PMID: 25735488]

[110] Chen S, Zhang Y, Geng H, Yang Y, Rui X, Li CC. Zinc ions pillared vanadate cathodes by chemical pre-intercalation towards long cycling life and low-temperature zinc ion batteries. J Power Sources 2019; 441: 227192.
[http://dx.doi.org/10.1016/j.jpowsour.2019.227192]

[111] Dong W, Shi JL, Wang TS, Yin YX, Wang CR, Guo YG. 3D zinc@carbon fiber composite framework anode for aqueous Zn–MnO$_2$ batteries. RSC Advances 2018; 8(34): 19157-63.
[http://dx.doi.org/10.1039/C8RA03226B] [PMID: 35539665]

[112] Fang G, Zhou J, Pan A, Liang S. Recent advances in aqueous zinc-ion batteries. ACS Energy Lett 2018; 3(10): 2480-501.
[http://dx.doi.org/10.1021/acsenergylett.8b01426]

[113] Gao XW, Deng YF, Wexler D, *et al.* Improving the electrochemical performance of the LiNi $_{0.5}$ Mn $_{1.5}$ O $_4$ spinel by polypyrrole coating as a cathode material for the lithium-ion battery. J Mater Chem A Mater Energy Sustain 2015; 3(1): 404-11.
[http://dx.doi.org/10.1039/C4TA04018J]

[114] Geng H, Cheng M, Wang B, Yang Y, Zhang Y, Li CC. Electronic structure regulation of layered vanadium oxide *via* interlayer doping strategy toward superior high-rate and low-temperature zinc-ion batteries. Adv Funct Mater 2020; 30(6): 1907684.
[http://dx.doi.org/10.1002/adfm.201907684]

[115] M. Ghosh, V. Vijayakumar, B. Anothumakkool, and S. Kurungot, Nafion ionomer-based single component electrolytes for aqueous Zn/MnO2 batteries with long cycle life, ACS Sustainable Chemistry & Engineering, 8, pp. 5040-5049, 2020.
[http://dx.doi.org/10.1021/acssuschemeng.9b06798]

[116] Ghosh M, Vijayakumar V, Kurungot S. Dendrite growth suppression by Zn $^{2+}$ -integrated nafion ionomer membranes: beyond porous separators toward aqueous Zn/V $_2$ O $_5$ batteries with extended cycle life. Energy Technol (Weinheim) 2019; 7(9): 1900442.
[http://dx.doi.org/10.1002/ente.201900442]

[117] Guo X, Li J, Jin X, *et al.* A hollow-structured manganese oxide cathode for stable zn-mno^2 batteries. Nanomaterials (Basel) 2018; 8(5): 301.
[http://dx.doi.org/10.3390/nano8050301] [PMID: 29734746]

[118] Hoang TKA, Doan TNL, Sun KEK, Chen P. Corrosion chemistry and protection of zinc & zinc alloys by polymer-containing materials for potential use in rechargeable aqueous batteries. RSC Advances 2015; 5(52): 41677-91.
[http://dx.doi.org/10.1039/C5RA00594A]

[119] Hu P, Zhu T, Wang X, *et al.* Highly durable Na $_2$ V $_6$ O $_{16}$ ·1.63h $_2$ o nanowire cathode for aqueous zinc-ion battery. Nano Lett 2018; 18(3): 1758-63.
[http://dx.doi.org/10.1021/acs.nanolett.7b04889] [PMID: 29397745]

[120] Huang M, Li F, Dong F, Zhang YX, Zhang LL. MnO $_2$ -based nanostructures for high-performance supercapacitors. J Mater Chem A Mater Energy Sustain 2015; 3(43): 21380-423.
[http://dx.doi.org/10.1039/C5TA05523G]

[121] Huang Y, Liu J, Huang Q, Zheng Z, Hiralal P, Zheng F, *et al.* Flexible high energy density zinc-ion batteries enabled by binder-free MnO^2/reduced graphene oxide electrode. Flexible Electron 2018; 2: 21.
[http://dx.doi.org/10.1038/s41528-018-0034-0]

[122] Huang Y, Mou J, Liu W, *et al.* Novel insights into energy storage mechanism of aqueous rechargeable zn/mno^2 batteries with participation of mn^{2-}. Nano-Micro Lett 2019; 11(1): 49.
[http://dx.doi.org/10.1007/s40820-019-0278-9] [PMID: 34138004]

[123] Islam S, Alfaruqi MH, Mathew V, *et al.* Facile synthesis and the exploration of the zinc storage mechanism of β-MnO $_2$ nanorods with exposed (101) planes as a novel cathode material for high performance eco-friendly zinc-ion batteries. J Mater Chem A Mater Energy Sustain 2017; 5(44): 23299-309.
[http://dx.doi.org/10.1039/C7TA07170A]

[124] Jeong SK, Seo HY, Kim DH, *et al.* Suppression of dendritic lithium formation by using concentrated electrolyte solutions. Electrochem Commun 2008; 10(4): 635-8.
[http://dx.doi.org/10.1016/j.elecom.2008.02.006]

[125] Jin Y, Zou L, Liu L, *et al.* Joint charge storage for high-rate aqueous zinc–manganese dioxide batteries. Adv Mater 2019; 31(29): 1900567.
[http://dx.doi.org/10.1002/adma.201900567] [PMID: 31157468]

[126] Kang L, Cui M, Jiang F, *et al.* Nanoporous caco $_3$ coatings enabled uniform zn stripping/plating for

long-life zinc rechargeable aqueous batteries. Adv Energy Mater 2018; 8(25): 1801090.
[http://dx.doi.org/10.1002/aenm.201801090]

[127] Khayum M A, Ghosh M, Vijayakumar V, *et al.* Zinc ion interactions in a two-dimensional covalent organic framework based aqueous zinc ion battery. Chem Sci (Camb) 2019; 10(38): 8889-94.
[http://dx.doi.org/10.1039/C9SC03052B] [PMID: 31762974]

[128] Lee B, Lee HR, Kim H, Chung KY, Cho BW, Oh SH. Elucidating the intercalation mechanism of zinc ions into α-MnO$_2$ for rechargeable zinc batteries. Chem Commun (Camb) 2015; 51(45): 9265-8.
[http://dx.doi.org/10.1039/C5CC02585K] [PMID: 25920416]

[129] Li H, Ma L, Han C, *et al.* Advanced rechargeable zinc-based batteries: Recent progress and future perspectives. Nano Energy 2019; 62: 550-87.
[http://dx.doi.org/10.1016/j.nanoen.2019.05.059]

[130] Li NW, Yin YX, Li JY, Zhang CH, Guo YG. Passivation of lithium metal anode *via* hybrid ionic liquid electrolyte toward stable li plating/Stripping. Adv Sci (Weinh) 2017; 4(2): 1600400.
[http://dx.doi.org/10.1002/advs.201600400] [PMID: 28251057]

[131] Li Y, Wang S, Salvador JR, *et al.* Reaction mechanisms for long-life rechargeable zn/mno$_2$ batteries. Chem Mater 2019; 31(6): 2036-47.
[http://dx.doi.org/10.1021/acs.chemmater.8b05093]

[132] Li Z, Wang J, Liu X, Liu S, Ou J, Yang S. Electrostatic layer-by-layer self-assembly multilayer films based on graphene and manganese dioxide sheets as novel electrode materials for supercapacitors. J Mater Chem 2011; 21(10): 3397.
[http://dx.doi.org/10.1039/c0jm02650f]

[133] Lindström H, Södergren S, Solbrand A, *et al.* Li$^+$ ion insertion in TiO$_2$ (anatase). 2. voltammetry on nanoporous films. J Phys Chem B 1997; 101(39): 7717-22.
[http://dx.doi.org/10.1021/jp970490q]

[134] Liu F, Chen Z, Fang G, *et al.* V2O5 Nanospheres with mixed vanadium valences as high electrochemically active aqueous zinc-ion battery cathode. Nano-Micro Lett 2019; 11(1): 25.
[http://dx.doi.org/10.1007/s40820-019-0256-2] [PMID: 34137986]

[135] Lu Y, Tikekar M, Mohanty R, Hendrickson K, Ma L, Archer LA. Stable cycling of lithium metal batteries using high transference number electrolytes. Adv Energy Mater 2015; 5(9): 1402073.
[http://dx.doi.org/10.1002/aenm.201402073]

[136] Ming J, Guo J, Xia C, Wang W, Alshareef H N. Zinc-ion batteries: Materials, mechanisms, and applications. Mater Sci Eng 2019; 135: 58.
[http://dx.doi.org/10.1016/j.mser.2018.10.002]

[137] Oberholzer P, Tervoort E, Bouzid A, Pasquarello A, Kundu D. Oxide *versus* nonoxide cathode materials for aqueous zn batteries: an insight into the charge storage mechanism and consequences thereof. ACS Appl Mater Interfaces 2019; 11(1): 674-82.
[http://dx.doi.org/10.1021/acsami.8b16284] [PMID: 30521309]

[138] Pan H, Shao Y, Yan P, *et al.* Reversible aqueous zinc/manganese oxide energy storage from conversion reactions. Nat Energy 2016; 1(5): 16039.
[http://dx.doi.org/10.1038/nenergy.2016.39]

[139] Poyraz AS, Laughlin J, Zec Z. Improving the cycle life of cryptomelane type manganese dioxides in aqueous rechargeable zinc ion batteries: The effect of electrolyte concentration. Electrochim Acta 2019; 305: 423-32.
[http://dx.doi.org/10.1016/j.electacta.2019.03.093]

[140] Qiu N, Chen H, Yang Z, Sun S, Wang Y. Synthesis of manganese-based complex as cathode material for aqueous rechargeable batteries. RSC Advances 2018; 8(28): 15703-8.
[http://dx.doi.org/10.1039/C8RA01982G] [PMID: 35539490]

[141] Selvakumaran D, Pan A, Liang S, Cao G. A review on recent developments and challenges of cathode

materials for rechargeable aqueous Zn-ion batteries. J Mater Chem A Mater Energy Sustain 2019; 7(31): 18209-36.
[http://dx.doi.org/10.1039/C9TA05053A]

[142] Shah DB, Olson KR, Karny A, Mecham SJ, DeSimone JM, Balsara NP. Effect of anion size on conductivity and transference number of perfluoroether electrolytes with lithium salts. J Electrochem Soc 2017; 164(14): A3511-7.
[http://dx.doi.org/10.1149/2.0301714jes]

[143] Song M, Tan H, Chao D, Fan HJ. Recent Advances in Zn-Ion Batteries. Adv Funct Mater 2018; 28(41): 1802564.
[http://dx.doi.org/10.1002/adfm.201802564]

[144] Stoševski I, Bonakdarpour A, Cuadra F, Wilkinson DP. Highly crystalline ramsdellite as a cathode material for near-neutral aqueous MnO $_2$ /Zn batteries. Chem Commun (Camb) 2019; 55(14): 2082-5.
[http://dx.doi.org/10.1039/C8CC07805J] [PMID: 30693914]

[145] Suo L, Borodin O, Gao T, *et al.* Water-in-salt electrolyte enables high-voltage aqueous lithium-ion chemistries. Science 2015; 350(6263): 938-43.
[http://dx.doi.org/10.1126/science.aab1595] [PMID: 26586759]

[146] Suo L, Borodin O, Wang Y, *et al.* Water-in-salt electrolyte makes aqueous sodium-ion battery safe, green, and long-lasting. Adv Energy Mater 2017; 7(21): 1701189.
[http://dx.doi.org/10.1002/aenm.201701189]

[147] Toupin M, Brousse T, Bélanger D. Charge storage mechanism of MnO $_2$ electrode used in aqueous electrochemical capacitor. Chem Mater 2004; 16(16): 3184-90.
[http://dx.doi.org/10.1021/cm049649j]

[148] Vijayakumar V, Diddens D, Heuer A, Kurungot S, Winter M, Nair JR. Dioxolanone-anchored poly(allyl ether)-based cross-linked dual-salt polymer electrolytes for high-voltage lithium metal batteries. ACS Appl Mater Interfaces 2020; 12(1): 567-79.
[http://dx.doi.org/10.1021/acsami.9b16348] [PMID: 31825198]

[149] Vijayakumar V, Ghosh M, Arun Torris AT, *et al.* Water-in-acid gel polymer electrolyte realized through a phosphoric acid-enriched polyelectrolyte matrix toward solid-state supercapacitors. ACS Sustain Chem& Eng 2018; 6: 12630.
[http://dx.doi.org/10.1021/acssuschemeng.8b01175]

[150] Wang J, Polleux J, Lim J, Dunn B. Pseudocapacitive contributions to electrochemical energy storage in tio $_2$ (anatase) nanoparticles. J Phys Chem C 2007; 111(40): 14925-31.
[http://dx.doi.org/10.1021/jp074464w]

[151] Wang JG, Yang Y, Huang ZH, Kang F. Coaxial carbon nanofibers/MnO^2 nanocomposites as freestanding electrodes for high-performance electrochemical capacitors. Electrochim Acta 2011; 56(25): 9240-7.
[http://dx.doi.org/10.1016/j.electacta.2011.07.140]

[152] Wang L, Cao X, Xu L, Chen J, Zheng J. Transformed akhtenskite MnO $_2$ from Mn $_3$ O $_4$ as cathode for a rechargeable aqueous zinc ion battery. ACS Sustain Chem& Eng 2018; 6(12): 16055-63.
[http://dx.doi.org/10.1021/acssuschemeng.8b02502]

[153] Wang LP, Li NW, Wang TS, Yin YX, Guo YG, Wang CR. Conductive graphite fiber as a stable host for zinc metal anodes. Electrochim Acta 2017; 244: 172-7.
[http://dx.doi.org/10.1016/j.electacta.2017.05.072]

[154] Wu B, Zhang G, Yan M, *et al.* Graphene scroll-coated α-MnO $_2$ nanowires as high-performance cathode materials for aqueous zn-ion battery. Small 2018; 14(13): 1703850.
[http://dx.doi.org/10.1002/smll.201703850]

[155] Wu X, Xiang Y, Peng Q, *et al.* Green-low-cost rechargeable aqueous zinc-ion batteries using hollow porous spinel ZnMn $_2$ O $_4$ as the cathode material. J Mater Chem A Mater Energy Sustain 2017; 5(34):

17990-7.
[http://dx.doi.org/10.1039/C7TA00100B]

[156]   Xu C, Li B, Du H, Kang F. Energetic zinc ion chemistry: the rechargeable zinc ion battery. Angew Chem Int Ed 2012; 51(4): 933-5.
[http://dx.doi.org/10.1002/anie.201106307] [PMID: 22170816]

[157]   Yang C, Chen J, Qing T, *et al.* 4.0 V Aqueous Li-Ion Batteries. Joule 2017; 1(1): 122-32.
[http://dx.doi.org/10.1016/j.joule.2017.08.009]

[158]   Yu P, Li C, Guo X. Sodium storage and pseudocapacitive charge in textured $Li_4Ti_5O_{12}$ thin films. J Phys Chem C 2014; 118(20): 10616-24.
[http://dx.doi.org/10.1021/jp5010693]

[159]   Zeng Y, Zhang X, Meng Y, *et al.* Achieving ultrahigh energy density and long durability in a flexible rechargeable quasi-solid-state zn-mno$_2$ battery. Adv Mater 2017; 29(26): 1700274.
[http://dx.doi.org/10.1002/adma.201700274]

[160]   Zeng Y, Zhang X, Qin R, *et al.* Dendrite-free zinc deposition induced by multifunctional cnt frameworks for stable flexible zn-ion batteries. Adv Mater 2019; 31(36): 1903675.
[http://dx.doi.org/10.1002/adma.201903675] [PMID: 31342572]

[161]   Zhan C, Lu J, Jeremy Kropf A, *et al.* Mn(II) deposition on anodes and its effects on capacity fade in spinel lithium manganate–carbon systems. Nat Commun 2013; 4(1): 2437.
[http://dx.doi.org/10.1038/ncomms3437] [PMID: 24077265]

[162]   Zhang L, Rodríguez-Pérez IA, Jiang H, *et al.* ZnCl$_2$ water-in-salt electrolyte transforms the performance of vanadium oxide as a zn battery cathode. Adv Funct Mater 2019; 29(30): 1902653.
[http://dx.doi.org/10.1002/adfm.201902653]

[163]   Zhang N, Cheng F, Liu J, *et al.* Rechargeable aqueous zinc-manganese dioxide batteries with high energy and power densities. Nat Commun 2017; 8(1): 405.
[http://dx.doi.org/10.1038/s41467-017-00467-x] [PMID: 28864823]

[164]   Zhang N, Cheng F, Liu Y, *et al.* Cation-deficient spinel $ZnMn_2O_4$ cathode in $Zn(CF_3SO_3)_2$ electrolyte for rechargeable aqueous zn-ion battery. J Am Chem Soc 2016; 138(39): 12894-901.
[http://dx.doi.org/10.1021/jacs.6b05958] [PMID: 27627103]

[165]   Zhao Q, Huang W, Luo Z, *et al.* High-capacity aqueous zinc batteries using sustainable quinone electrodes. Sci Adv 2018; 4: eaao1761.
[http://dx.doi.org/10.1126/sciadv.aao1761]

[166]   Zhao S, Han B, Zhang D, *et al.* Unravelling the reaction chemistry and degradation mechanism in aqueous Zn/MnO$_2$ rechargeable batteries. J Mater Chem A Mater Energy Sustain 2018; 6(14): 5733-9.
[http://dx.doi.org/10.1039/C8TA01031E]

[167]   Tanaka Y, Tanaka Y. Infrared spectra of bisarylsulfonimide derivatives. Chem Pharm Bull (Tokyo) 1974; 22(11): 2546-51.
[http://dx.doi.org/10.1248/cpb.22.2546]

[168]   Tarascon J M, Gozdz A S, Schmutz C, Shokoohi F, Warren P C. Performance of bellcore's plastic rechargeable li-ion batteries. Solid State Ionics 1996; 86(Part 1): 49-54.
[http://dx.doi.org/10.1016/0167-2738(96)00330-X]

[169]   Moers O, Henschel D, Lange I, Blaschette A, Jones PG. Onium-di(arensulfonyl)amide: Von der gestreckten zur gefalteten Konformation des $(ArSO^2)2N\theta$-Anions. Z Anorg Allg Chem 2000; 626(11): 2388-98.
[http://dx.doi.org/10.1002/1521-3749(200011)626:11<2388::AID-ZAAC2388>3.0.CO;2-I]

[170]   Wiers B M, Foo M-L, Balsara N P, Long J R. A solid lithium electrolyte *via* addition of lithium isopropoxide to a metal–organic framework with open metal sites. Journal of the American Chemical Society 2011; 133: 14522-5.
[http://dx.doi.org/10.1021/ja205827z]

[171] Capiglia C, Saito Y, Kataoka H, Kodama T, Quartarone E, Mustarelli P. Structure and transport properties of polymer gel electrolytes based on PVdF-HFP and LiN(C2F5SO^2)2. Solid State Ion 2000; 131(3-4): 291-9.
       [http://dx.doi.org/10.1016/S0167-2738(00)00678-0]

[172] Yeon S-H, Kim K-S, Choi S, Cha J-H, Lee H. Characterization of PVdF(HFP) Gel Electrolytes Based on 1-(2-Hydroxyethyl)-3-methyl Imidazolium Ionic Liquids. The Journal of Physical Chemistry B 2005; 109: 17928-35.
       [http://dx.doi.org/10.1021/jp053237w]

[173] Zhang Y, Ting J, Rohan R, *et al.* Fabrication of a proton exchange membrane *via* blended sulfonimide functionalized polyamide. Journal of Materials Science 2014; 49: 3442-50.
       [http://dx.doi.org/10.1007/s10853-014-8055-0]

[174] Liang Y, Ji L, Guo B, *et al.* Preparation and electrochemical characterization of ionic-conducting lithium lanthanum titanate oxide/polyacrylonitrile submicron composite fiber-based lithium-ion battery separators. J Power Sources 2011; 196(1): 436-41.
       [http://dx.doi.org/10.1016/j.jpowsour.2010.06.088]

[175] Zhu Y S, Gao X W, Wang X J, Hou Y Y, Liu L L, Wu Y P. A single-ion polymer electrolyte based on boronate for lithium ion batteries. Electrochemistry Communications 2012; 22: 29-32.
       [http://dx.doi.org/10.1016/j.elecom.2012.05.022]

[176] Zhu Y, Xiao S, Shi Y, Yang Y, Hou Y, Wu Y. A composite gel polymer electrolyte with high performance based on poly(vinylidene fluoride) and polyborate for lithium ion batteries. Advanced Energy Materials 2014; 4.
       [http://dx.doi.org/10.1002/aenm.201300647]

[177] Zhu Y, Wang F, Liu L, Xiao S, Yang Y, Wu Y. Cheap glass fiber mats as a matrix of gel polymer electrolytes for lithium ion batteries. Sci Rep 2013; 3.
       [http://dx.doi.org/10.1038/srep03187]

[178] Zhu Y, Wang F, Liu L, Xiao S, Chang Z, Wu Y. Composite of a nonwoven fabric with poly(vinylidene fluoride) as a gel membrane of high safety for lithium ion battery. Energy Environ Sci 2013; 6(2): 618-24.
       [http://dx.doi.org/10.1039/C2EE23564A]

[179] Meziane R, Bonnet JP, Courty M, Djellab K, Armand M. Single-ion polymer electrolytes based on a delocalized polyanion for lithium batteries. Electrochim Acta 2011; 57: 14-9.
       [http://dx.doi.org/10.1016/j.electacta.2011.03.074]

[180] Cai W, Zhang Y, Li J, Sun Y, Cheng H. Single-ion polymer electrolyte membranes enable lithium-ion batteries with a broad operating temperature range. ChemSusChem 2014; 7(4): 1063-7.
       [http://dx.doi.org/10.1002/cssc.201301373] [PMID: 24623577]

[181] Zhu Y, Xiao S, Shi Y, Yang Y, Wu Y. A trilayer poly(vinylidene fluoride)/polyborate/poly(vinylidene fluoride) gel polymer electrolyte with good performance for lithium ion batteries. J Mater Chem A Mater Energy Sustain 2013; 1(26): 7790-7.
       [http://dx.doi.org/10.1039/c3ta00167a]

[182] K. Sinha, W. Wang, K. I. Winey, and J. K. Maranas, Dynamic Patterning in PEO-Based Single Ion Conductors for Li Ion Batteries, Macromolecules, 45, pp. 4354-4362, 2012.
       [http://dx.doi.org/10.1021/ma300051y]

[183] Matsumoto K, Endo T. Synthesis of networked polymers by copolymerization of monoepoxy-substituted lithium sulfonylimide and diepoxy-substituted poly(ethylene glycol), and their properties. J Polym Sci A Polym Chem 2011; 49(8): 1874-80.
       [http://dx.doi.org/10.1002/pola.24614]

[184] Chazalviel JN. Electrochemical aspects of the generation of ramified metallic electrodeposits. Phys Rev A 1990; 42(12): 7355-67.

[http://dx.doi.org/10.1103/PhysRevA.42.7355] [PMID: 9904050]

[185] Du Pasquier A, Warren PC, Culver D, Gozdz AS, Amatucci GG, Tarascon JM. Plastic PVDF-HFP electrolyte laminates prepared by a phase-inversion process. Solid State Ion 2000; 135(1-4): 249-57.
[http://dx.doi.org/10.1016/S0167-2738(00)00371-4]

[186] Li G, Li Z, Zhang P, Zhang H, Wu Y. Research on a gel polymer electrolyte for Li-ion batteries. Pure Appl Chem 2008; 80(11): 2553-63.
[http://dx.doi.org/10.1351/pac200880112553]

[187] Watanabe M, Kanba M, Matsuda H, *et al.* High lithium ionic conductivity of polymeric solid electrolytes. Makromol Chem, Rapid Commun 1981; 2(12): 741-4.
[http://dx.doi.org/10.1002/marc.1981.030021208]

[188] Choe H S, Giaccai J, Alamgir M, Abraham K M. Preparation and characterization of poly(vinyl sulfone)- and poly(vinylidene fluoride)-based electrolytes. Electrochimica Acta 1995; 40: 2289-93.
[http://dx.doi.org/10.1016/0013-4686(95)00180-M]

[189] Cui ZY, Du CH, Xu YY, Ji GL, Zhu BK. Preparation of porous PVdF membrane *via* thermally induced phase separation using sulfolane. J Appl Polym Sci 2008; 108(1): 272-80.
[http://dx.doi.org/10.1002/app.27494]

[190] Z.-Y. Cui, Y.-Y. Xu, L.-P. Zhu, J.-Y. Wang, and B.-K. Zhu, Investigation on PVDF-HFP microporous membranes prepared by TIPS process and their application as polymer electrolytes for lithium ion batteries, Ionics, 15, pp. 469-476, 2009.
[http://dx.doi.org/10.1007/s11581-008-0253-9]

[191] Appetecchi G B, Croce F, Scrosati B. Kinetics and stability of the lithium electrode in poly(methylmethacrylate)-based gel electrolytes. Electrochimica Acta 1995; 40: 991-7.
[http://dx.doi.org/10.1016/0013-4686(94)00345-2]

[192] Sekhon S. Solvent effect on gel electrolytes containing lithium salts. Solid State Ion 2000; 136-137(-2): 1189-92.
[http://dx.doi.org/10.1016/S0167-2738(00)00584-1]

[193] Deepa M, Sharma N, Agnihotry S A, Singh S, Lal T, Chandra R. Conductivity and viscosity of liquid and gel electrolytes based on LiClO4, LiN(CF3SO2)2 and PMMA. Solid State Ionics 2002; 253-8.
[http://dx.doi.org/10.1016/S0167-2738(02)00307-7]

[194] Singh B, Kumar R, Sekhon SS. Conductivity and viscosity behaviour of PMMA based gels and nano dispersed gels: Role of dielectric constant of the solvent. Solid State Ion 2005; 176(17-18): 1577-83.
[http://dx.doi.org/10.1016/j.ssi.2005.05.001]

[195] Y. Huai, J. Gao, Z. Deng, and J. Suo, Preparation and characterization of a special structural poly(acrylonitrile)-based microporous membrane for lithium-ion batteries, Ionics, 16, pp. 603-611, 2010.
[http://dx.doi.org/10.1007/s11581-010-0431-4]

[196] Pu W, He X, Wang L, Tian Z, Jiang C, Wan C. Preparation of P(AN–MMA) microporous membrane for Li-ion batteries by phase inversion. J Membr Sci 2006; 280(1-2): 6-9.
[http://dx.doi.org/10.1016/j.memsci.2006.05.028]

[197] Akashi H, Tanaka K, Sekai K. A flexible Li polymer primary cell with a novel gel electrolyte based on poly(acrylonitrile). J Power Sources 2002; 104(2): 241-7.
[http://dx.doi.org/10.1016/S0378-7753(01)00966-1]

[198] Akashi H, Shibuya M, Orui K, Shibamoto G, Sekai K. Practical performances of Li-ion polymer batteries with LiNi0.8Co0.2O^2, MCMB, and PAN-based gel electrolyte. J Power Sources 2002; 112(2): 577-82.
[http://dx.doi.org/10.1016/S0378-7753(02)00465-2]

[199] Rajendran S, Babu RS, Sivakumar P. Effect of salt concentration on poly (vinyl chloride)/poly (acrylonitrile) based hybrid polymer electrolytes. J Power Sources 2007; 170(2): 460-4.

[http://dx.doi.org/10.1016/j.jpowsour.2007.04.041]

[200] Perera K S, Dissanayake M A K L, Skaarup S, West K. Application of polyacrylonitrile-based polymer electrolytes in rechargeable lithium batteries. Journal of Solid State Electrochemistry 2008; 12: 873-7.
[http://dx.doi.org/10.1007/s10008-007-0479-x]

[201] Ramesh S, Arof A K. Structural, thermal and electrochemical cell characteristics of poly(vinyl chloride)-based polymer electrolytes. Journal of Power Sources 2001; 99: 41-7.
[http://dx.doi.org/10.1016/S0378-7753(00)00690-X]

[202] Tian Z, Pu W, He X, Wan C, Jiang C. Preparation of a microporous polymer electrolyte based on poly(vinyl chloride)/poly(acrylonitrile-butyl acrylate) blend for Li-ion batteries. Electrochim Acta 2007; 52(9): 3199-206.
[http://dx.doi.org/10.1016/j.electacta.2006.09.068]

[203] Rhoo H-J, Kim H-T, Park J-K, Hwang T-S. Ionic conduction in plasticized PVCPMMA blend polymer electrolytes 1997; 42: 1571-9.

[204] Bonardelli P, Moggi G, Turturro A. Glass transition temperatures of copolymer and terpolymer fluoroelastomers. Polymer 1986; 27: 905-9.
[http://dx.doi.org/10.1016/0032-3861(86)90302-2]

[205] Elmér AM, Jannasch P. Polymer electrolyte membranes by *in situ* polymerization of poly(ethylene carbonate-co-ethylene oxide) macromonomers in blends with poly(vinylidene fluoride-co-hexafluoropropylene). J Polym Sci, B, Polym Phys 2007; 45(1): 79-90.
[http://dx.doi.org/10.1002/polb.20980]

[206] Hardy L C, Shriver D F. Preparation and electrical response of solid polymer electrolytes with only one mobile species. Journal of the American Chemical Society 1985; 107: 3823-8.
[http://dx.doi.org/10.1021/ja00299a012]

[207] Idris NH, Rahman MM, Wang JZ, Liu HK. Microporous gel polymer electrolytes for lithium rechargeable battery application. J Power Sources 2012; 201: 294-300.
[http://dx.doi.org/10.1016/j.jpowsour.2011.10.141]

[208] Sun Y, Rohan R, Cai W, *et al.* A polyamide single-ion electrolyte membrane for application in lithium-ion batteries. Energy Technol (Weinheim) 2014; 2(8): 698-704.
[http://dx.doi.org/10.1002/ente.201402041]

[209] Song JY, Wang YY, Wan CC. Review of gel-type polymer electrolytes for lithium-ion batteries. J Power Sources 1999; 77(2): 183-97.
[http://dx.doi.org/10.1016/S0378-7753(98)00193-1]

[210] Sun XG, Kerr JB. Synthesis and characterization of network single ion conductors based on comb-branched polyepoxide ethers and lithium bis(allylmalonato)borate. Macromolecules 2006; 39(1): 362-72.
[http://dx.doi.org/10.1021/ma0507701]

[211] Sun XG, Reeder CL, Kerr JB. Synthesis and characterization of network type single ion conductors. Macromolecules 2004; 37(6): 2219-27.
[http://dx.doi.org/10.1021/ma035690g]

[212] Croce F, Sacchetti S, Scrosati B. Advanced, lithium batteries based on high-performance composite polymer electrolytes. J Power Sources 2006; 162(1): 685-9.
[http://dx.doi.org/10.1016/j.jpowsour.2006.07.038]

[213] Sadoway D R, Huang B, Trapa P E, Soo P P, Bannerjee P, Mayes A M. Self-doped block copolymer electrolytes for solid-state, rechargeable lithium batteries. Journal of Power Sources 2001; 97-98: 621-3.
[http://dx.doi.org/10.1016/S0378-7753(01)00642-5]

[214] Zhang Y, Rohan R, Cai W, *et al.* Influence of chemical microstructure of single-ion polymeric

electrolyte membranes on performance of lithium-ion batteries. ACS Applied Materials & Interfaces 2014; 6: 17534-42.
[http://dx.doi.org/10.1021/am503152m]

[215] Zhao J, Wang L, He X, Wan C, Jiang C. Determination of lithium-ion transference numbers in lipf6–pc solutions based on electrochemical polarization and nmr measurements. J Elec Soc 2008; 155: A292-6.

[216] Raghavan P, Zhao X, Kim JK, *et al.* Ionic conductivity and electrochemical properties of nanocomposite polymer electrolytes based on electrospun poly(vinylidene fluoride-co-hexafluoropropylene) with nano-sized ceramic fillers. Electrochim Acta 2008; 54(2): 228-34.
[http://dx.doi.org/10.1016/j.electacta.2008.08.007]

[217] Yasin S M M, Ibrahim S, Johan M R. Effect of zirconium oxide nanofiller and dibutyl phthalate plasticizer on ionic conductivity and optical properties of solid polymer electrolyte. The Scientific World Journal 2014; 2014: 547076.
[http://dx.doi.org/10.1155/2014/547076]

[218] Mandal BK, Walsh CJ, Sooksimuang T, *et al.* New class of single-ion-conducting solid polymer electrolytes derived from polyphenols. Chemistry of Materials 2000; 12: 6-8.
[http://dx.doi.org/10.1021/cm9906497]

[219] Wilson J, Ravi G, Kulandainathan MA. Electrochemical studies on inert filler incorporated poly (vinylidene fluoride - hexafluoropropylene) (PVDF - HFP) composite electrolytes. Polímeros 2006; 16(2): 88-93.
[http://dx.doi.org/10.1590/S0104-14282006000200006]

[220] Malmonge LF, Malmonge JA, Sakamoto WK. Study of pyroelectric activity of PZT/PVDF-HFP composite. Mater Res 2003; 6(4): 469-73.
[http://dx.doi.org/10.1590/S1516-14392003000400007]

[221] Ladouceur S, Paillet S, Vijh A, Guerfi A, Dontigny M, Zaghib K. Synthesis and characterization of a new family of aryl-trifluoromethanesulfonylimide Li-Salts for Li-ion batteries and beyond. J Power Sources 2015; 293: 78-88.
[http://dx.doi.org/10.1016/j.jpowsour.2015.05.011]

[222] Bloch E, Phan T, Bertin D, Llewellyn P, Hornebecq V. Direct synthesis of mesoporous silica presenting large and tunable pores using BAB triblock copolymers: Influence of each copolymer block on the porous structure. Microporous Mesoporous Mater 2008; 112(1-3): 612-20.
[http://dx.doi.org/10.1016/j.micromeso.2007.10.051]

[223] LIU ZhiHong QB, HaiYan Y, YuLong Duan, JianJun Zhang. The research progress of single-ion conducting polymer lithium borate salts. Scientia Sinica Chimica 2014; 44: 1229-40.

[224] Bouchet R, Maria S, Meziane R, *et al.* Single-ion BAB triblock copolymers as highly efficient electrolytes for lithium-metal batteries. Nat Mater 2013; 12: 452-7.
[http://dx.doi.org/10.1038/nmat3602]

[225] Feng S, Shi D, Liu F, *et al.* Single lithium-ion conducting polymer electrolytes based on poly[(4-styrenesulfonyl)(trifluoromethanesulfonyl)imide] anions. Electrochim Acta 2013; 93: 254-63.
[http://dx.doi.org/10.1016/j.electacta.2013.01.119]

[226] Ma ZH, Han HB, Zhou ZB, Nie J. SBA-15-supported poly(4-styrenesulfonyl(perfluorobutylsulfonyl)imide) as heterogeneous Brønsted acid catalyst for synthesis of diindolylmethane derivatives. J Mol Catal Chem 2009; 311(1-2): 46-53.
[http://dx.doi.org/10.1016/j.molcata.2009.06.021]

[227] Sandí G, Carrado KA, Joachin H, Lu W, Prakash J. Polymer nanocomposites for lithium battery applications. J Power Sources 2003; 119-121: 492-6.
[http://dx.doi.org/10.1016/S0378-7753(03)00272-6]

[228] Oh H, Xu K, Yoo HD, *et al.* Poly(arylene ether)-based single-ion conductors for lithium-ion batteries.

Chem Mater 2015.

[229]   Liu Y, Zhang Y, Pan M, *et al.* A mechanically robust porous single ion conducting electrolyte membrane fabricated *via* self-assembly. Journal of Membrane Science
[http://dx.doi.org/10.1016/j.memsci.2016.02.002]

[230]   Ma Q, Zhang H, Zhou C, *et al.* Single lithium-ion conducting polymer electrolytes based on a super-delocalized polyanion. Angew Chem Int Ed 2016; 55(7): 2521-5.
[http://dx.doi.org/10.1002/anie.201509299] [PMID: 26840215]

[231]   Van Humbeck JF, Aubrey ML, Alsbaiee A, *et al.* Tetraarylborate polymer networks as single-ion conducting solid electrolytes. Chem Sci (Camb) 2015; 6(10): 5499-505.
[http://dx.doi.org/10.1039/C5SC02052B] [PMID: 28757947]

[232]   Wang L, Li N, He X, Wan C, Jiang C. Macromolecule plasticized interpenetrating structure solid state polymer electrolyte for lithium ion batteries. Electrochim Acta 2012; 68: 214-9.
[http://dx.doi.org/10.1016/j.electacta.2012.02.067]

[233]   Watanabe M, Suzuki Y, Nishimoto A. Single ion conduction in polyether electrolytes alloyed with lithium salt of a perfluorinated polyimide. Electrochim Acta 2000; 45(8-9): 1187-92.
[http://dx.doi.org/10.1016/S0013-4686(99)00380-1]

[234]   Tatsuma T, Taguchi M, Oyama N. Inhibition effect of covalently cross-linked gel electrolytes on lithium dendrite formation. Electrochim Acta 2001; 46(8): 1201-5.
[http://dx.doi.org/10.1016/S0013-4686(00)00706-4]

[235]   Shin W-K, Kannan AG, Kim D-W. Effective suppression of dendritic lithium growth using an ultrathin coating of nitrogen and sulfur codoped graphene nanosheets on polymer separator for lithium metal batteries. ACS Applied Materials & Interfaces 2015; 7: 23700-7.
[http://dx.doi.org/10.1021/acsami.5b07730]

[236]   Wu H, Zhuo D, Kong D, Cui Y. Improving battery safety by early detection of internal shorting with a bifunctional separator. Nat Commun 2014; 5.
[http://dx.doi.org/10.1038/ncomms6193]

[237]   Lee YS, Lee JH, Choi JA, Yoon WY, Kim DW. Cycling characteristics of lithium powder polymer batteries assembled with composite gel polymer electrolytes and lithium powder anode. Adv Funct Mater 2013; 23(8): 1019-27.
[http://dx.doi.org/10.1002/adfm.201200692]

[238]   Andersson AM, Edström K. Chemical composition and morphology of the elevated temperature sei on graphite. J Electrochem Soc 2001; 148(10): A1100.
[http://dx.doi.org/10.1149/1.1397771]

[239]   Aurbach D, Markovsky B, Talyossef Y, Salitra G, Kim HJ, Choi S. Studies of cycling behavior, ageing, and interfacial reactions of LiNi0.5Mn1.5O4 and carbon electrodes for lithium-ion 5-V cells. J Power Sources 2006; 162(2): 780-9.
[http://dx.doi.org/10.1016/j.jpowsour.2005.07.009]

[240]   Aurbach D, Talyosef Y, Markovsky B, *et al.* Design of electrolyte solutions for Li and Li-ion batteries: a review. Electrochim Acta 2004; 50(2-3): 247-54.
[http://dx.doi.org/10.1016/j.electacta.2004.01.090]

[241]   Carroll KJ, Yang MC, Veith GM, Dudney NJ, Meng YS. Intrinsic surface stability in limn2−xnixo4−δ (x = 0.45, 0.5) high voltage spinel materials for lithium ion batteries. Electrochem Solid-State Lett 2012; 15(5): A72.
[http://dx.doi.org/10.1149/2.008206esl]

[242]   Chen Z, Lu WQ, Liu J, Amine K. LiPF6/LiBOB blend salt electrolyte for high-power lithium-ion batteries. Electrochim Acta 2006; 51(16): 3322-6.
[http://dx.doi.org/10.1016/j.electacta.2005.09.027]

[243]   Dalavi S, Xu M, Knight B, Lucht BL. Effect of added libob on high voltage (lini0.5mn1.5o4) spinel

cathodes. Electrochem Solid-State Lett 2011; 15(2): A28-31.
[http://dx.doi.org/10.1149/2.015202esl]

[244] Demeaux J, Caillon-Caravanier M, Galiano H, Lemordant D, Claude-Montigny B. LiNi $_{0.4}$ Mn $_{1.6}$ O $_4$ /electrolyte and carbon black/electrolyte high voltage interfaces: to evidence the chemical and electronic contributions of the solvent on the cathode-electrolyte interface formation. J Electrochem Soc 2012; 159(11): A1880-90.
[http://dx.doi.org/10.1149/2.052211jes]

[245] Duncan H, Duguay D, Abu-Lebdeh Y, Davidson IJ. Study of the LiMn $_{1.5}$ Ni $_{0.5}$ O $_4$ /Electrolyte interface at room temperature and 60°C. J Electrochem Soc 2011; 158(5): A537-45.
[http://dx.doi.org/10.1149/1.3567954]

[246] Goodenough JB, Kim Y. Challenges for rechargeable li batteries. Chem Mater 2010; 22(3): 587-603.
[http://dx.doi.org/10.1021/cm901452z]

[247] Ha SY, Han JG, Song YM, *et al.* Using a lithium bis(oxalato) borate additive to improve electrochemical performance of high-voltage spinel LiNi0.5Mn1.5O4 cathodes at 60°C. Electrochim Acta 2013; 104: 170-7.
[http://dx.doi.org/10.1016/j.electacta.2013.04.082]

[248] Jiang J, Dahn JR. Comparison of the thermal stability of lithiated graphite in LiBOB EC/DEC and in LiPF[sub 6] EC/DEC. Electrochem Solid-State Lett 2003; 6(9): A180.
[http://dx.doi.org/10.1149/1.1592911]

[249] Julien CM, Mauger A. Review of 5-V electrodes for Li-ion batteries: status and trends. Ionics 2013; 19(7): 951-88.
[http://dx.doi.org/10.1007/s11581-013-0913-2]

[250] Kawamura T, Kimura A, Egashira M, Okada S, Yamaki JI. Thermal stability of alkyl carbonate mixed-solvent electrolytes for lithium ion cells. J Power Sources 2002; 104(2): 260-4.
[http://dx.doi.org/10.1016/S0378-7753(01)00960-0]

[251] Kim JH, Pieczonka NPW, Li Z, Wu Y, Harris S, Powell BR. Understanding the capacity fading mechanism in LiNi0.5Mn1.5O4/graphite Li-ion batteries. Electrochim Acta 2013; 90: 556-62.
[http://dx.doi.org/10.1016/j.electacta.2012.12.069]

[252] La Mantia F, Huggins RA, Cui Y. Oxidation processes on conducting carbon additives for lithium-ion batteries. J Appl Electrochem 2013; 43(1): 1-7.
[http://dx.doi.org/10.1007/s10800-012-0499-9]

[253] Liu J, Manthiram A. Understanding the improved electrochemical performances of fe-substituted 5 v spinel cathode LiMn $_{1.5}$ Ni $_{0.5}$ O $_4$. J Phys Chem C 2009; 113(33): 15073-9.
[http://dx.doi.org/10.1021/jp904276t]

[254] Lu D, Xu M, Zhou L, Garsuch A, Lucht BL. Failure mechanism of graphite/LiNi $_{0.5}$ Mn $_{1.5}$ O $_4$ Cells at High Voltage and Elevated temperature. J Electrochem Soc 2013; 160(5): A3138-43.
[http://dx.doi.org/10.1149/2.022305jes]

[255] Lux SF, Lucas IT, Pollak E, Passerini S, Winter M, Kostecki R. The mechanism of HF formation in LiPF6 based organic carbonate electrolytes. Electrochem Commun 2012; 14(1): 47-50.
[http://dx.doi.org/10.1016/j.elecom.2011.10.026]

[256] MahootcheianAsl N, Kim J-H, Pieczonka NPW, Liu Z, Kim Y. Multilayer electrolyte cell: A new tool for identifying electrochemical performances of high voltage cathode materials. Electrochem Commun 2013; 32: 1-4.
[http://dx.doi.org/10.1016/j.elecom.2013.03.031]

[257] Matsui M, Dokko K, Akita Y, Munakata H, Kanamura K. Surface layer formation of LiCoO^2 thin film electrodes in non-aqueous electrolyte containing lithium bis(oxalate)borate. J Power Sources 2012; 210: 60-6.
[http://dx.doi.org/10.1016/j.jpowsour.2012.02.042]

[258]   Panitz JC, Wietelmann U, Wachtler M, Ströbele S, Wohlfahrt-Mehrens M. Film formation in LiBOB-containing electrolytes. J Power Sources 2006; 153(2): 396-401.
[http://dx.doi.org/10.1016/j.jpowsour.2005.05.025]

[259]   Patoux S, Daniel L, Bourbon C, *et al.* High voltage spinel oxides for Li-ion batteries: From the material research to the application. J Power Sources 2009; 189(1): 344-52.
[http://dx.doi.org/10.1016/j.jpowsour.2008.08.043]

[260]   Pieczonka NPW, Liu Z, Lu P, *et al.* Understanding transition-metal dissolution behavior in $LiNi_{0.5}Mn_{1.5}O_4$ high-voltage spinel for lithium ion batteries. J Phys Chem C 2013; 117(31): 15947-57.
[http://dx.doi.org/10.1021/jp405158m]

[261]   Pieczonka N P W, Yang L, Balogh M P, *et al.* Impact of lithium bis(oxalate)borate electrolyte additive on the performance of high-voltage spinel/graphite li-ion batteries. The Journal of Physical Chemistry C 2013; 117: 22603-12.
[http://dx.doi.org/10.1021/jp408717x]

[262]   Schweiger HG, Multerer M, Schweizer-Berberich M, Gores HJ. Optimization of cycling behavior of lithium ion cells at 60 °c by additives for electrolytes based on lithium bis[1,2-oxalato(2-)-o, o] borate. Int J Electrochem Sci 2008; 3: 427.

[263]   Shin DW, Bridges CA, Huq A, Paranthaman MP, Manthiram A. Role of cation ordering and surface segregation in high-voltage spinel $LiMn_{1.5}Ni_{0.5-x}M_xO_4$ (M = Cr, Fe, and Ga) cathodes for lithium-ion batteries. Chem Mater 2012; 24(19): 3720-31.
[http://dx.doi.org/10.1021/cm301844w]

[264]   Sloop SE, Kerr JB, Kinoshita K. The role of Li-ion battery electrolyte reactivity in performance decline and self-discharge. J Power Sources 2003; 119-121: 330-7.
[http://dx.doi.org/10.1016/S0378-7753(03)00149-6]

[265]   Xiao A, Yang L, Lucht BL. Thermal reactions of lipf[sub 6] with added LiBOB. Electrochem Solid-State Lett 2007; 10(11): A241.
[http://dx.doi.org/10.1149/1.2772084]

[266]   Xu K. Nonaqueous liquid electrolytes for lithium-based rechargeable batteries. Chem Rev 2004; 104(10): 4303-418.
[http://dx.doi.org/10.1021/cr030203g] [PMID: 15669157]

[267]   Xu K. Tailoring electrolyte composition for LiBOB. J Electrochem Soc 2008; 155(10): A733.
[http://dx.doi.org/10.1149/1.2961055]

[268]   Xu K, Lee U, Zhang S, Wood M, Jow TR. Chemical analysis of graphite/electrolyte interface formed in libob-based electrolytes. Electrochem Solid-State Lett 2003; 6(7): A144.
[http://dx.doi.org/10.1149/1.1576049]

[269]   Xu W, Angell CA. Weakly coordinating anions, and the exceptional conductivity of their nonaqueous solutions. Electrochem Solid-State Lett 2001; 4(1): E1.
[http://dx.doi.org/10.1149/1.1344281]

[270]   Xu W, Shusterman AJ, Marzke R, Angell CA. LiMOB, an unsymmetrical nonaromatic orthoborate salt for nonaqueous solution electrochemical applications. J Electrochem Soc 2004; 151(4): A632.
[http://dx.doi.org/10.1149/1.1651528]

[271]   Xu W, Shusterman AJ, Videa M, Velikov V, Marzke R, Angell CA. Structures of orthoborate anions and physical properties of their lithium salt nonaqueous solutions. J Electrochem Soc 2003; 150(1): E74.
[http://dx.doi.org/10.1149/1.1527939]

[272]   Yang L, Ravdel B, Lucht BL. Electrolyte reactions with the surface of high voltage lini[sub 0.5]mn[sub 1.5]o[sub 4] cathodes for lithium-ion batteries. Electrochem Solid-State Lett 2010; 13(8): A95.
[http://dx.doi.org/10.1149/1.3428515]

[273] Zhang SS, Xu K, Jow TR. Enhanced performance of Li-ion cell with LiBF4-PC based electrolyte by addition of small amount of LiBOB. J Power Sources 2006; 156(2): 629-33.
[http://dx.doi.org/10.1016/j.jpowsour.2005.04.023]

[274] Zhang SS. A review on electrolyte additives for lithium-ion batteries. J Power Sources 2006; 162(2): 1379-94.
[http://dx.doi.org/10.1016/j.jpowsour.2006.07.074]

[275] Zhang X, Devine TM. Passivation of aluminum in lithium-ion battery electrolytes with LiBOB. J Electrochem Soc 2006; 153(9): B365.
[http://dx.doi.org/10.1149/1.2218269]

[276] Zheng J, Xiao J, Xu W, *et al.* Surface and structural stabilities of carbon additives in high voltage lithium ion batteries. J Power Sources 2013; 227: 211-7.
[http://dx.doi.org/10.1016/j.jpowsour.2012.11.038]

[277] Zhuang GV, Xu K, Jow TR, Ross PN. Study of sei layer formed on graphite anodes in pc/libob electrolyte using ir spectroscopy. Electrochem Solid-State Lett 2004; 7(8): A224.
[http://dx.doi.org/10.1149/1.1756855]

[278] Zhang Y, Sun Y, Xu G, *et al.* Lithium-ion batteries with a wide temperature range operability enabled by highly conductive sp^3 boron-based single ion polymer electrolytes. Energy Technol (Weinheim) 2014; 2(7): 643-50.
[http://dx.doi.org/10.1002/ente.201402010]

[279] Xu G, Sun Y, Rohan R, Zhang Y, Cai W, Cheng H. A lithium poly(pyromellitic acid borate) gel electrolyte membrane for lithium-ion batteries. J Mater Sci 2014; 49(17): 6111-7.
[http://dx.doi.org/10.1007/s10853-014-8341-x]

[280] Rohan R, Pareek K, Cai W, *et al.* Melamine–terephthalaldehyde–lithium complex: a porous organic network based single ion electrolyte for lithium ion batteries. J Mater Chem A Mater Energy Sustain 2015; 3(9): 5132-9.
[http://dx.doi.org/10.1039/C4TA06855F]

[281] Rohan R, Pareek K, Chen Z, *et al.* A high performance polysiloxane-based single ion conducting polymeric electrolyte membrane for application in lithium ion batteries. J Mater Chem A Mater Energy Sustain 2015; 3(40): 20267-76.
[http://dx.doi.org/10.1039/C5TA02628H]

[282] Zhang Y, Xu G, Sun Y, Han B. A class of sp3 boron-based single-ion polymeric electrolytes for lithium ion batteries. RSC Advances 2013; 3: 14934-7.

[283] Cai W, Li J, Zhang Y, *et al.* Current status and future prospects of research on single ion polymer electrolyte for lithium battery applications. Journal of The Chinese Ceramic Society 2014; 1: 78-92.

[284] Deng K, Zeng Q, Wang D, *et al.* Single-ion conducting gel polymer electrolytes: design, preparation and application. J Mater Chem A Mater Energy Sustain 2020; 8(4): 1557-77.
[http://dx.doi.org/10.1039/C9TA11178F]

[285] Rohan R, Kuo TC, Chen MW, Lee JT. Nanofiber single-ion conducting electrolytes: an approach for high-performance lithium batteries at ambient temperature. ChemElectroChem 2017; 4(9): 2178-83.
[http://dx.doi.org/10.1002/celc.201700389]

[286] Miller JR, Simon P. Materials science. Electrochemical capacitors for energy management. Science 2008; 321(5889): 651-2.
[http://dx.doi.org/10.1126/science.1158736] [PMID: 18669852]

[287] Pandolfo A G, Hollenkamp A F. Carbon properties and their role in supercapacitors. Journal of Power Sources 2006; 157: 11-27.
[http://dx.doi.org/10.1016/j.jpowsour.2006.02.065]

[288] Chang L, Hang Hu Y. 2.21 SupercapacitorsComprehensive Energy Systems. Elsevier 2018; pp. 663-

95.
[http://dx.doi.org/10.1016/B978-0-12-809597-3.00247-9]

[289] Xu B, Yue S, Sui Z, *et al.* What is the choice for supercapacitors: graphene or graphene oxide? Energy Environ Sci 2011; 4(8): 2826-30.
[http://dx.doi.org/10.1039/c1ee01198g]

[290] Zhou R, Meng C, Zhu F, *et al.* High-performance supercapacitors using a nanoporous current collector made from super-aligned carbon nanotubes. Nanotechnology 2010; 21: 345701.
[http://dx.doi.org/10.1088/0957-4484/21/34/345701]

[291] Meng C, Liu C, Fan S. Flexible carbon nanotube/polyaniline paper-like films and their enhanced electrochemical properties. Electrochem Commun 2009; 11(1): 186-9.
[http://dx.doi.org/10.1016/j.elecom.2008.11.005]

[292] Mirmohseni A, Wallace GG. Preparation and characterization of processable electroactive polyaniline–polyvinyl alcohol composite. Polymer (Guildf) 2003; 44(12): 3523-8.
[http://dx.doi.org/10.1016/S0032-3861(03)00242-8]

[293] Prasad KR, Munichandraiah N. Electrochemical studies of polyaniline in a gel polymer electrolyte, Electrochem. Solid-State 2002; 5: A271.

[294] Pushparaj VL, Shaijumon MM, Kumar A, *et al.* Flexible energy storage devices based on nanocomposite paper. Proc Natl Acad Sci USA 2007; 104(34): 13574-7.
[http://dx.doi.org/10.1073/pnas.0706508104] [PMID: 17699622]

[295] Simon P, Gogotsi Y. Materials for electrochemical capacitors. Nat Mater 2008; 7(11): 845-54.
[http://dx.doi.org/10.1038/nmat2297] [PMID: 18956000]

[296] Sivaraman P, Kushwaha RK, Shashidhara K, *et al.* All solid supercapacitor based on polyaniline and crosslinked sulfonated poly[ether ether ketone]. Electrochim Acta 2010; 55(7): 2451-6.
[http://dx.doi.org/10.1016/j.electacta.2009.12.009]

[297] Sivaraman P, Rath SK, Hande VR, Thakur AP, Patri M, Samui AB. All-solid-supercapacitor based on polyaniline and sulfonated polymers. Synth Met 2006; 156(16-17): 1057-64.
[http://dx.doi.org/10.1016/j.synthmet.2006.06.017]

[298] Staiti P, Lufrano F. A study of the electrochemical behaviour of electrodes in operating solid-state supercapacitors. Electrochim Acta 2007; 53(2): 710-9.
[http://dx.doi.org/10.1016/j.electacta.2007.07.039]

[299] Subramania A, Kalyana Sundaram NT, Vijaya Kumar G, Vasudevan T. New polymer electrolyte based on (PVA–PAN) blend for Li-ion battery applications. Ionics 2006; 12(2): 175-8.
[http://dx.doi.org/10.1007/s11581-006-0018-2]

[300] Wang D, Song P, Liu C, Wu W, Fan S. Highly oriented carbon nanotube papers made of aligned carbon nanotubes. Nanotechnology 2008; 19(7): 075609.
[http://dx.doi.org/10.1088/0957-4484/19/7/075609] [PMID: 21817646]

[301] Wang DW, Li F, Zhao J, *et al.* Fabrication of graphene/polyaniline composite paper *via in situ* anodic electropolymerization for high-performance flexible electrode. ACS Nano 2009; 3(7): 1745-52.
[http://dx.doi.org/10.1021/nn900297m] [PMID: 19489559]

[302] Xiao-Feng WANG, Dian-Bo RUAN, Da-Zhi WANG, Ji LIANG. Hybrid electrochemical supercapacitors based on polyaniline and activated carbon electrodes. Wuli Huaxue Xuebao 2005; 21(3): 261-6.
[http://dx.doi.org/10.3866/PKU.WHXB20050307]

[303] Wang Y, Zhang XG. All solid-state supercapacitor with phosphotungstic acid as the proton-conducting electrolyte. Solid State Ion 2004; 166(1-2): 61-7.
[http://dx.doi.org/10.1016/j.ssi.2003.11.001]

[304] Wu Q, Xu Y, Yao Z, Liu A, Shi G. Supercapacitors based on flexible graphene/polyaniline nanofiber

composite films. ACS Nano 2010; 4(4): 1963-70.
[http://dx.doi.org/10.1021/nn1000035] [PMID: 20355733]

[305] Xiao L, Chen Z, Feng C, *et al.* Flexible, stretchable, transparent carbon nanotube thin film loudspeakers. Nano Lett 2008; 8(12): 4539-45.
[http://dx.doi.org/10.1021/nl802750z] [PMID: 19367976]

[306] Zhou R, Meng C, Zhu F, *et al.* High-performance supercapacitors using a nanoporous current collector made from super-aligned carbon nanotubes. Nanotechnology 2010; 21(34): 345701.
[http://dx.doi.org/10.1088/0957-4484/21/34/345701] [PMID: 20683140]

[307] Zhong C, Deng Y, Hu W, Qiao J, Zhang L, Zhang J. A review of electrolyte materials and compositions for electrochemical supercapacitors. Chem Soc Rev 2015; 44(21): 7484-539.
[http://dx.doi.org/10.1039/C5CS00303B] [PMID: 26050756]

[308] Tsay K-C, Zhang L, Zhang J. Effects of electrode layer composition/thickness and electrolyte concentration on both specific capacitance and energy density of supercapacitor. Electrochimica Acta 2012; 60: 428-36.
[http://dx.doi.org/10.1016/j.electacta.2011.11.087]

[309] Komaba S, Hasegawa T, Dahbi M, Kubota K. Potassium intercalation into graphite to realize high-voltage/high-power potassium-ion batteries and potassium-ion capacitors. Electrochemistry Communications 2015; 60: 172-5.
[http://dx.doi.org/10.1016/j.elecom.2015.09.002]

[310] Armand M, Endres F, MacFarlane DR, Ohno H, Scrosati B. Ionic-liquid materials for the electrochemical challenges of the future. Nat Mater 2009; 8(8): 621-9.
[http://dx.doi.org/10.1038/nmat2448] [PMID: 19629083]

[311] Ashby M, Shercliff H, Cebon D. Materials Engineering, Science, Processing and Design. Elsevier 2007.

[312] Azaïs P, Duclaux L, Florian P, *et al.* Causes of supercapacitors ageing in organic electrolyte. J Power Sources 2007; 171(2): 1046-53.
[http://dx.doi.org/10.1016/j.jpowsour.2007.07.001]

[313] Balducci A, Dugas R, Taberna PL, *et al.* High temperature carbon–carbon supercapacitor using ionic liquid as electrolyte. J Power Sources 2007; 165(2): 922-7.
[http://dx.doi.org/10.1016/j.jpowsour.2006.12.048]

[314] Feng G, Cummings PT. Supercapacitor capacitance exhibits oscillatory behavior as a function of nanopore size. J Phys Chem Lett 2011; 2(22): 2859-64.
[http://dx.doi.org/10.1021/jz201312e]

[315] Frackowiak E. Carbon materials for supercapacitor application. Phys Chem Chem Phys 2007; 9(15): 1774-85.
[http://dx.doi.org/10.1039/b618139m] [PMID: 17415488]

[316] Jänes A, Thomberg T, Kurig H, Lust E. Nanoscale fine-tuning of porosity of carbide-derived carbon prepared from molybdenum carbide. Carbon 2009; 47(1): 23-9.
[http://dx.doi.org/10.1016/j.carbon.2008.07.010]

[317] Kondrat S, Georgi N, Fedorov MV, Kornyshev AA. A superionic state in nano-porous double-layer capacitors: insights from Monte Carlo simulations. Phys Chem Chem Phys 2011; 13(23): 11359-66.
[http://dx.doi.org/10.1039/c1cp20798a] [PMID: 21566824]

[318] Lin R, Taberna P-L, Fantini S, *et al.* Capacitive energy storage from −50 °C to 100 °C using an ionic liquid electrolyte. J Phys Chem Lett 2011; 2(19): 2396-401.
[http://dx.doi.org/10.1021/jz201065t]

[319] Miller JR, Burke A. Electrochemical capacitors: Challenges and opportunities for real-world applications. Electrochem Soc Interface 2008; 17(1): 53-7.
[http://dx.doi.org/10.1149/2.F08081IF]

[320] Miller JR, Outlaw RA, Holloway BC. Graphene double-layer capacitor with ac line-filtering performance. Science 2010; 329(5999): 1637-9.
[http://dx.doi.org/10.1126/science.1194372] [PMID: 20929845]

[321] Portet C, Yushin G, Gogotsi Y. Electrochemical performance of carbon onions, nanodiamonds, carbon black and multiwalled nanotubes in electrical double layer capacitors. Carbon 2007; 45(13): 2511-8.
[http://dx.doi.org/10.1016/j.carbon.2007.08.024]

[322] Raymundo-Piñero E, Kierzek K, Machnikowski J, Béguin F. Relationship between the nanoporous texture of activated carbons and their capacitance properties in different electrolytes. Carbon 2006; 44(12): 2498-507.
[http://dx.doi.org/10.1016/j.carbon.2006.05.022]

[323] Shim Y, Kim HJ. Solvation of carbon nanotubes in a room-temperature ionic liquid. ACS Nano 2009; 3(7): 1693-702.
[http://dx.doi.org/10.1021/nn900195b] [PMID: 19583191]

[324] Simon P, Gogotsi Y. Capacitive energy storage in nanostructured carbon–electrolyte systems. Accounts of Chemical Research 2013; 46: 1094-103.
[http://dx.doi.org/10.1021/ar200306b]

[325] Tsuda T, Hussey CL. Electrochemical applications of room-temperature ionic liquids. Electrochem Soc Interface 2007; 16(1): 42-9.
[http://dx.doi.org/10.1149/2.F05071IF]

[326] Zhai Y, Dou Y, Zhao D, Fulvio PF, Mayes RT, Dai S. Carbon materials for chemical capacitive energy storage. Adv Mater 2011; 23(42): 4828-50.
[http://dx.doi.org/10.1002/adma.201100984] [PMID: 21953940]

[327] Zhao X, Sánchez BM, Dobson PJ, Grant PS. The role of nanomaterials in redox-based supercapacitors for next generation energy storage devices. Nanoscale 2011; 3(3): 839-55.
[http://dx.doi.org/10.1039/c0nr00594k] [PMID: 21253650]

[328] Zhu Y, Murali S, Stoller MD, *et al.* Carbon-based supercapacitors produced by activation of graphene. Science 2011; 332(6037): 1537-41.
[http://dx.doi.org/10.1126/science.1200770] [PMID: 21566159]

[329] Suriyakumar S, Bhardwaj P, Grace AN, Stephan AM. Role of polymers in enhancing the performance of electrochemical supercapacitors: a review. Batter Supercaps 2021; 4(4): 571-84.
[http://dx.doi.org/10.1002/batt.202000272]

[330] Lian K, Li J, Virya A, Wu H. The impact of polymer electrolytes on the performance and longevity of solid flexible supercapacitors. 2019 IEEE International Flexible Electronics Technology Conference (IFETC) 2019; 1-2.
[http://dx.doi.org/10.1109/IFETC46817.2019.9073759]

CHAPTER 5

Polymer Composite Membrane for Microbial Fuel Cell Application

Kalpana Sharma[1], Anusha Vempaty[1], Barun Kumar[1], Shweta Rai[1], Vaibhav Raj[1], Deepak Jadhav[2] and Soumya Pandit[1,*]

[1] *Department of Life Sciences, School of Basic Sciences and Research, Sharda University, Greater Noida – 201306, India*

[2] *Department of Agricultural Engineering, Maharashtra Institute of Technology, Aurangabad-431010, India*

Abstract: Energy production is a demanded process in today's world. Some processes might generate pollutants and other undesirable particulates and toxic chemicals. One such eco-friendly and efficient method for generating electricity and energy can be through fuel cells with the utilization of microbes (bacteria). Such a method can be termed Microbial Fuel Cells (MFCs). It is a bio-electrochemical system. It uses bacteria and their biochemical processes for generating an electric current, along with oxygen which is a high-energy oxidant. MFCs imitate the bacterial interactions that are found in the nature. Being a cell, it requires electrodes, substrates, and electrolytic solutions. To improve the efficiency of the MFC, we need to separate the anode and cathode into two compartments and the respective reactions taking place. Membranes play a crucial role in achieving it. A membrane not only divides the anode from the cathode but also prevents the entry of oxygen into the anode chamber. The most important function of a membrane is to allow the selective transfer of ions across the two electrode chambers. Membranes can be diaphragms or separators. Porous membranes are commercially used ones usually made of different effective polymer materials. Other important membranes can be semi-permeable and ion-exchange membranes. This chapter mainly reviews the various membranes and the materials used in their structures that have the potential to increase the MFC performance. It also focuses on the different transport processes across the membranes, along with a brief of advances in this technology and future scope.

Keywords: Ion-exchange, Mass transfer, Materials, Membranes, Microbial fuel cell, Porous separators.

---

* **Corresponding author Soumya Pandit:** Department of Life Sciences, School of Basic Sciences and Research, Sharda University, Greater Noida – 201306, India; E-mail: soumya.pandit@sharda.ac.in

Subhendu Bhandari, Prashant Gupta and Ayan Dey (Eds.)

INTRODUCTION

To meet the tremendous need for energy supply for the huge population, an environment-friendly and sustainable form of production is crucial. It should also be economically stable by simultaneously maintaining the ecological balance during storage and conversion [1]. A much familiar form of energy is fossil fuels – coal and petroleum. However, these compounds are non-renewable, and their combustion leads to pollution in the surroundings, ultimately causing global warming. Hence, a more ecological and sustainable alternative is required for efficient energy production [2].

A fuel cell is the most dependable decision with proficient energy change innovation which is responsible for the conversion of chemical energy into electrical energy through electrochemical responses specifically anodic oxidation and cathodic reduction responses [3]. These cells are spotless energy change gadgets, where the oxidant and reductant are constantly provided to generate power, unlike batteries that contain pre-stuffed chemical constituents. They can give long-haul arrangements as sustainable and effective energy transformation gadgets with the least or zero discharge of greenhouse gases [4]. Huge ecological advantages are normal on fuel cells, especially for the automobile sector and energy production for fixed and mobile applications.

A fuel cell is an electrochemical cell, which consistently changes the chemical energy of a fuel and an oxidant to electrical energy in a cathode-electrolyte framework, intended for constant nourishment of reactants at a high temperature of an electrocatalyst to catalyze the oxidation and reduction responses [5].

Microbial Fuel Cells (MFCs) are the bio-electrochemical systems that use bacteria for the production of electric current. The biochemical reactions are catalyzed by bacteria. When we see the general construction of an MFC, we find electrodes, wirings, a salt bridge, and a membrane. The membranes play a crucial role in the cell [6].

A membrane is usually any medium that is used to separate the electrodes and hence, to obtain two different compartments. The flow of substances through these compartments gets minimized when they encounter the membrane [7]. This is to check and keep a balance between different chemicals, ions, or substances across the cell [8]. Even in living beings, all the cells have cell membranes that separate the interior of all cells from the outside environment. Hence, we understand the importance of membranes [9].

The main characteristics of the membrane material are:

- Pore size
- Porosity
- Capacity of ion-exchange

Membranes can be of two types:

- Permeable – also called diaphragms. They allow the flow of liquid as a whole (less filtering of substances passing through) and hence, it is non-selective in the transportation of ions or molecules.
- Semipermeable –mostly depends on the size of the molecules that pass through as well as the charge possessed. It allows selective transport of certain species or substances (ions or molecules).

Porous separators or membranes are the most commonly used separators. Their structures have pores of the size around 1–50 μm (Fig. **1**) which prevents the mixing of gaseous products and solid particles. Throughout the cell, ionic motion carries a current that results in a voltage drop, ΔV, across the separator. It is expressed in terms of the effective specific resistance, ρ (Ω cm), of the electrolyte in the material:

$$\Delta V = IR = I(\rho L / A) = (I / A)L / \kappa \tag{1}$$

or

$$\Delta V = j\rho L \tag{2}$$

Here, R - resistance, j - current density (A cm^{-2}), κ - conductivity (S cm^{-1}), ρL - area-specific resistance

Ionic and material transport across the membrane is through diffusion. In some cases, a small net flow is essential to prevent the back transport of certain species (in opposite direction). Hence, we can say that the membranes separate the anode and the cathode and make two electrolyte compartments. The flow of electrolyte thus is from one compartment to the other through the membrane. The performance of the membrane depends on its ability to control the transport of substances through its structure.

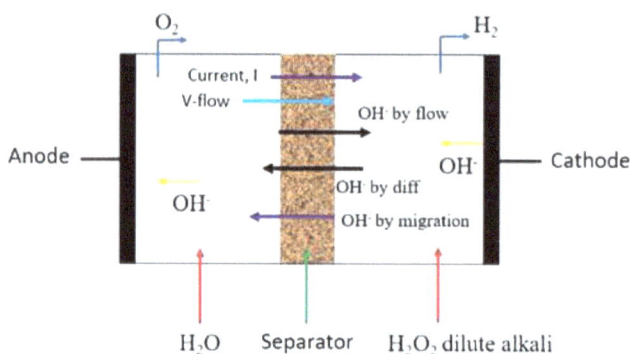

Fig. (1). Ion transfer in separators.

TYPES OF FUEL CELLS

Fuel cells are categorized as direct and indirect based on working temperature, the condition of matter of the components, and the sort of electrolyte utilized. Based on working temperature, fuel cells can be further distinguished as low-temperature and high-temperature cells.

Proton exchange membrane fuel cell (PEMFC), direct methanol fuel cell (DMFC), and alkaline fuel cell (AFC) are set under the classification of low-temperature cells, whereas solid oxide fuel cell (SOFC), molten carbonate fuel cell (MCFC), and phosphoric acid fuel cell (PAFC) are located under high-temperature cell units.

Among this arrangement, PEMFCs are more alluring and solid than other cells because of the wide scope of use, high effectiveness, and zero outflows of toxins [10].

Proton Exchange Membrane Fuel Cell (PEMFC)

The proton exchange membrane fuel cell is also known as the polymer electrolyte membrane fuel cell. This type of cell operates below 100^0C with extraordinary polymer electrolyte layers. This low-temperature fuel cell is the favoured decision for vehicles, convenient applications like handheld gadgets due to its fast start-up, low working temperature, and astounding energy productivity. The electrolyte in this system is an ion-exchange membrane – perfluorosulfonic acid (PFSA) polymer or Nafion [11]. The solitary result in this energy unit is water, and the maintenance of water in the PEMFC is basic for the effective execution. The fuel cell should work under the conditions where the resultant water does not sublimate quickly than it is delivered because the layer ought to be held under hydrated conditions.

Oxygen reduction reaction (ORR) is a multistep mechanism including electron transport with the establishment of various intermediates relying upon the pH of the electrolyte [12]. In PEMFC, the ORR can continue by two distinctive response pathways as follows [13]:

- Direct four-electron transport leading to water development as the solitary result.
- Two-electron transport step leading to the generation of H_2O_2 transition product that further goes through a reduction in a progression of electron transport to form water.

The degree of reduction seems to rely upon the selected catalyst material. It affirms the job of the electrocatalyst on the effectiveness of fuel cells, which incorporates the decision of catalyst and its supporting material. The 4-electron step is the most preferred pathway since it delivers a high voltage for an H_2-O_2 PEM energy cell [14].

CELL SEPARATORS

Separators are usually adopted in any electrochemical energy production due to their tendency to reduce the chemical reactions taking place between cathode and anode. The efficiency of a separator depends upon its inert nature, mechanical strength, good ionic conductivity, and being a non-conductor in the case of electricity. MFC uses separators to minimize energy losses due to oxidation and reduction.

Diaphragms and Porous Polymer Membranes

A porous membrane/diaphragm allows a limited flow of ions across it through diffusion. The pores should be uniformly distributed throughout the structure. The more are the number of pores or the larger is the size of pores, the electrical conductivity will be greater. But this will also lead to poor separation of the two electrodes from each other (when the main role of a membrane is to give good separation between the electrodes). If the material is thin, then the transport across the porous membrane will be more and the concentration gradient will be high. The membrane material must be non-conductive of electricity. Otherwise, the membrane will also start acting as an electrode due to the voltage gradient across the electrolyte. Many inorganic materials can be used for porous membrane-like ceramics, metal oxides, and asbestos [15]. There are certain drawbacks of using these materials, and hence, these are not suitable for making porous membranes. Polymeric materials like ethylene, propylene, vinyl chloride, and tetrafluoroethylene are much more efficient for the membrane than inorganic materials. The porous structure is incorporated at the time of fabrication during manufacturing. In case of MECs, during the evolution of hydrogen gas, the product gases need to be separated. Here, the membranes used should be hydrophilic in nature. Fluoropolymer membranes need suitable wetting agents like ZrO_2, to make them hydrophilic. Zirfon is a very good example of that. Porex and Celgard are some of the porous polymers that are used commonly as cell separators.

Semipermeable Membranes: Ion-Exchange Membranes

Ion-exchange membranes are those membranes that are specific to the passage of either anions or cations through them, as shown in Fig (**2**). Hence, they are selective. Low electrical resistance, high permeability and selectivity, good mechanical and dimensional stability, good chemical stability, and its capability of operating over a wide range of current densities and under varying conditions of temperature, current density, pH *etc* are the very essential properties of an efficient ion-exchange membrane. These properties are determined by two parameters: (i) basic polymer matrix, and (ii) the type and concentration of fixed ionic group. The basic polymer matrix is responsible for the mechanical, chemical, and thermal stability, whereas the type and concentration of the fixed ionic group determines the electrical properties and the selectivity of the membrane. There are two types of ion-exchange membranes:

- Cation-exchange membranes (CEMs) – allow only cations to pass through and are made from strong or weak acids. Sulfonic acid and carboxylic acid groups are usually used to make CEM.
- Anion-exchange membranes (AEMs) – allow the passage of only anions and are made of strong or weak bases. Quaternary ammonium group, tertiary, secondary, primary amines, and phosphonium and sulfonium groups are used to make AEM.

CEM is more stable than AEM in electrochemical processes.

Table **1** displays the various functional groups, either cationic or anionic which can be incorporated into the polymer matrix.

ION EXCHANGE MEMBRANE FOR THE CONVENTIONAL PRODUCTION PROCESS

Electrodialysis has been one of the main monetarily accessible large-scale water desalination measures dependent on membranes. In recent times, various electrodialysis-related cycles with bipolar membranes have been designed. These cycles are monetarily accessible and generally utilized in the treatment of certain industrial effluents, in the generation of ultrapure water, or deionization of food and drug products. Bilayer ion-exchange membranes are utilized in distinguishing a single charged ion from a multiple charged one. Electrodialysis through an ion-exchange membrane is used for treating wastewater while conserving the environment and recuperating the resources. Electrochemical characteristics of charged resins present in these membranes are collaborated with the membrane permeability to treat effluents and desalination of water. One such technique is designated as Membrane Capacitive Deionisation (MCDI). Thermoelectric conversion of power is also possible using thesemembranes. In this paper, we mainly discuss the two types of ion-exchange membranes and their performance.

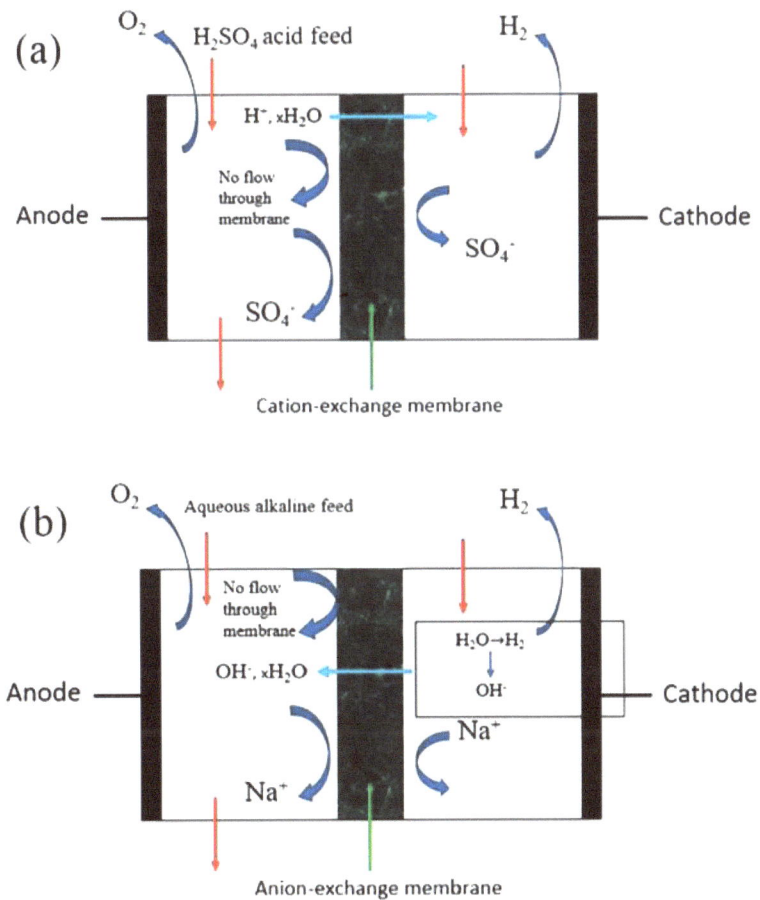

Fig. (2). Selective transfer of ion-exchange membranes. (**a**) Cation exchange membrane behavior. (**b**) Anion exchange membrane behavior.

Table 1. Cationic and anionic groups incorporated into membranes.

Ions	Functional Groups	Property	References
Cation	Sulfonate, $-SO_3$, $-H^+$	Strongly acidic	[16]
-	Carboxylate, -COOH, Phosphonate	Weakly acidic	[16]
Anion	Quaternary ammonium, $-N(CH_3)_3^+$ OH^-, Tri-ethylammonium, Trialkylbenzyl ammonium	Strongly basic	[16]
-	Trimethylamine, $-N(CH_3)_2$ Triethylamine, DABCO, Dimethyl sulfide, CH_3-S-CH_3, Diethyl sulfide, N,N,N', N'-Tetramethyl-1,6-hexane diamine	Weakly basic	[16]

MEMBRANES IN MFC

Till now, we have come to understand the concept of membranes and their importance. So, we understand that a membrane is usually used in electrochemical cells to prevent the transfer of oxygen into an anode compartment along with the motive of separation of the two electrodes from each other. It is designed in a way to selectively permit the flow of ions through it. In MFCs, ion-exchange membranes are used instead of others. It is of two types, each specific to the passage of either anion or cation, referred to as ion-exchange membranes. The different characteristic features of ion-exchange membranes are listed in Table **2**.

Table 2. Characteristics of commercial ion-exchange membranes.

Ion-exchange Membranes	Type	Perm-selectivity	Conductivity (10^4 s/m^2)	Power Density (w/m^2)	Ref.
AEM			-		
Neosepta AFN[16]	Homogenous	88.9	1.43	1.23-1.30	[17]
Fumasep FAD[31]	Homogenous	86	1.12	1.16-1.24	[17]
Ralex AMH-PES[16]	Homogenous	89.3	0.13	0.73-1.12	[17]
CEM			-		
Neosepta CM-1[31]	Homogenous	97.2	0.60	1.12-1.30	[17]
Fumasep FKD[16]	Homogenous	89.5	0.47	0.99-1.19	[17]
Ralex CMH-PES[16]	Homogenous	94.7	0.09	0.73-1.23	[17]

Cation-Exchange Membranes

The most commonly used polymer material for the membrane is perfluorosulfonic acid (PFSA). Such membranes are also called ionomer membranes. The PFSA structure consists of three regions:

- Polytetrafluoroethylene (PTFE) backbone
- Vinyl ethers containing side chains
- Clusters of sulfonic acid groups

Mobile counter-ions (cations) balance the negatively charged sulfonate groups (fixed charge) to attain electrical neutrality in the cell. Electrostatic forces tend to eliminate the mobile co-ions (anions) due to the similarity in their charge with that of the fixed charge groups. This process of elimination of co-ions is called Donnan exclusion and hence, CEM effectively enables the selective passage of cations. The clustered sulfonate group is hydrophilic and hence, imparts the property of ion-conductivity to the membrane. Due to this reason, the cations gain mobility when the membrane becomes hydrated. Hence, a potential gradient is created allowing free flow for H^+ ions. Nafion, Flemion, and Aquivion are the manufacturer names under which PFSA membranes are made.

Equivalent weight and ion-exchange capacity are the necessary features of the ion-exchange membranes [18]. If the membranes are thin, they tend to have less mechanical strength and higher reactant crossover (*e.g.*, oxygen diffusion). This will lead to a decrease in the utilization of fuel, and will cause polarization of the electrode. Peroxy groups are formed which corrodes the material. On the other hand, thick membranes will reduce the reactant crossover [19]. This also has the disadvantage of increasing the resistance, leading to low power density and efficiency. Other than PFSA membranes, nonfluorinated membranes, hydrocarbon membranes, partially sulfonated polyarylenes [20], and composite membranes using inert inorganic materials are being tested for use [21].

Anion-Exchange Membranes

AEMs generally contain positive ionic groups (poly-N^+Me_3) and mobile negatively charged anions (OH^-). The main characteristics of AEMs are the faster catalysis of fuel cell reactions, lower activation losses, less corrosiveness, and hence, cheap materials can be utilized for the cell components. AEMs can use non-metal and carbon catalysts. In water management, AEMs play a great role, especially in crossover and cathode flooding. This is because the transport of water and ion happens from cathode to anode. In most of the media, the OH^- ions tend to have a lower diffusion coefficient than that of protons. Hence, to achieve a good conductivity, the ion exchange capacities can be increased [22].

Factors for using AEMs in MFCs:

- OH^- ions react with CO_2 to form carbonate/bicarbonate.
- Solid metal carbonate precipitates (Na_2CO_3 or K_2CO_3) are formed. Deposition of

these precipitates will hinder the pores that are filled with electrolytes in the electrodes. This will result in the disruption of catalyst functioning and will hinder the membrane. AEMs have no mobile cation (Na^+ or K^+) which will lead to the precipitation of solid crystals, as the cations are already immobilized on the polymer.

Mostly, AEM is not stable in an alkaline medium. This is because the ammonium group is displaced by nucleophilic anions (OH^-). This can happen in two ways:

- Direct nucleophilic displacement.
- If β-hydrogens are present, then Hofmann elimination takes place.

Tertiary amines and methanol are formed when OH^- ions displace methyl (CH_3) groups.

Alkaline pH causes degradation of the membrane due to the attack of OH^- ions on the bonding of the ion-exchange functional group [23]. This disadvantage can be eliminated by using carbonate ions. Many other membranes are being tested and used to tackle the problem of instability in alkaline conditions. Radiation grafting produces much cheaper ionomer membranes [24]. Vinylbenzyl chloride-grafted fluorinated films along with amination leads to the production of alkaline anion exchange membranes (AAEMs) [25]. These membranes can be chloromethylated polysulfone while most of these membranes contain either trimethylammonium or N-methyl pyridinium groups [26].

Bipolar Membranes

Bipolar membranes (BPMs) are the derivatives of ion-exchange membranes comprised by a cation- and an anion-exchange layer, permitting the evolution of protons and hydroxide ions by employing a water separation component, as featured in Fig (**3**). Such a characteristic allows bipolar layers as appealing for a wide range of applications in numerous areas, biochemical industry, food production, and energy change and capacity. Bipolar films constitute mechanical support that is positioned between a cation exchange layer (CEL) and an anion exchange layer (AEL). CEL is shaped by the cation exchange scattering on one side of the mechanical support. AEL is shaped from the utilization of an anion exchange scattering on the contrary side of the mechanical support. In the middle layer, water is separated into OH^- and H^+ particles while surpassing a difference of 0.8 V in the potential of the system. Splitting of water and electrodialysis are the two significant utilizations of BPM.

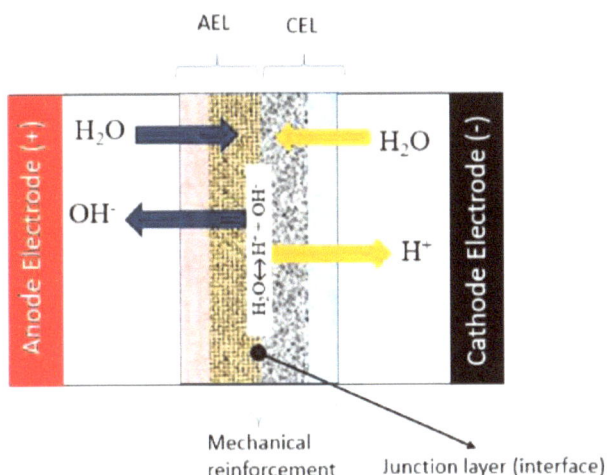

Fig. (3). Bipolar Membranes.

Membrane Requirements in MFCs

Certain factors need to be considered while using membranes in MFC:

- Oxygen has to be prevented from entering the anode chamber due to the anaerobic nature of the anode. But membranes will function only if it is wet and because of this oxygen diffuses into the membrane structure. Hence, it becomes a difficult factor to control. This problem can be eliminated by applying differential pressure from the anode to the cathode. This will stop the flow of oxygen.
- CEMs like PFSAs transport other cations (Na^+, K^+, NH_4^+, Ca_2^+, Mg_2^+) found in the anolyte in MFCs. Concentrations of these cations are 105 times higher than proton concentration in wastewater. Hence, the other cations are transported more readily through the membrane leading to the precipitation of salt on the cathode. This will inhibit the cathode catalyst.
- The formation of a biofilm is a common phenomenon during long-term operations, that deteriorates the effective functioning of MFC. To get rid of these biofilms (formed on the surface or inside the structure), certain chemical cleaning agents are used. Hence, the membrane must be chemical resistant.
- pH values in the anode chamber range around 5-9 which is also similar to the pH range inside the cathode chamber. The oxidation-reduction reaction at such a pH is unknown for most of the materials. Hence, there is a possibility of the production of hydrogen peroxide from this reaction.

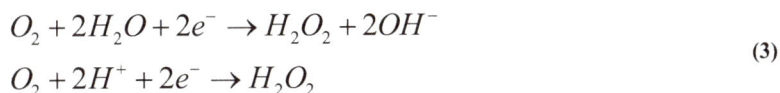

$$O_2 + 2H_2O + 2e^- \rightarrow H_2O_2 + 2OH^-$$
$$O_2 + 2H^+ + 2e^- \rightarrow H_2O_2$$

(3)

- Before its usage, a membrane is always pre-treated to enhance the transfer of appropriate cation (H^+ ions) or anion. However, it comes back to its functional form on exposure to another cation (K^+).
- Nonionic species (organic substrates, CO_2, O_2, organic compounds, electrolyte salts) can be transported through diffusion in ion-exchange membranes.
- Nonionic species should not be allowed to move to the cathode as it would affect the reduction of oxygen. This can lead to decrease in the efficiency of the oxidation-reduction reaction. As a result, polarization increases, the cathode surface is contaminated and the catalytic activity is reduced.

TRANSPORT PROCESS IN MEMBRANES

Ion Transport Processes

In MFCs, the mechanism involving the transfer of ions from the anode to cathode is as follows:

- Convection - this happens when the electrolyte is in motion, leading to the transfer of ions.
- Electric migration – this happens when an electrical potential gradient is responsible for the transfer of ions .
- Diffusion - this happens when a chemical potential gradient (concentration) is responsible for the transfer of ions .

Ion flux balances the charge difference that occurs between anode and cathode.

Nernst–Planck equation defines the flux densities, N [27]:

$$N_i = -D_i \frac{dC_i}{dx} + \frac{z_i F}{RT} D_i C_i (d\Phi / dx) + C_i U \qquad (4)$$

Here, the first term of the equation - diffusional mass transport, the second term – migration, the third term - convection ion flow under a velocity U, D_i - diffusion coefficient of ion species i, C_i - species concentration, F - Faraday's constant, R - gas constant, T – temperature, Φ - electric field, Z_i - charge on ion i .

Diffusional transport is defined by MacMullin number, NM, as a function of the tortuosity and porosity:

$$D_{eff} = D_i / N_M \qquad (5)$$

Here, D_{eff} - effective ionic diffusion coefficient

The required overall flux is given by:

$$N_i = -D_{eff}\frac{dC_i}{dx} + \frac{z_i F}{RT}D_{eff}C_i(d\Phi/dx) + C_{i,m}U \tag{6}$$

The current density and potential gradient are related to each other by the equation:

$$j = -\kappa\frac{d\Phi}{dx} \tag{7}$$

Here, κ - conductivity

This conductivity is described by:

$$\kappa = \frac{F^2}{RT}\sum D_i C_i z_i^2 \tag{8}$$

Hence, the flux in the membrane becomes:

$$N_i = -D_{eff}\frac{dC_i}{dx} - j\frac{z_i F}{\kappa RT}D_{eff}C_i + C_{i,m}U \tag{9}$$

Ion-Exchange Membranes and the Transport of Ions

Current density is defined as the movement of ions in an electrolyte solution by:

$$j = F\sum z_i N_i \tag{10}$$

Transport number, t_i is defined as the ratio of the partial current density associated with one ion to the total current density:

$$t_i = \frac{j_i}{j} = \frac{F^2}{RT} \frac{D_i C_i z_i^2}{\kappa} \tag{11}$$

When convection is negligible, the net flow in the solution across the membrane is zero. Hence, the flux will be:

$$N_i = -D_i \frac{dC_i}{dx} - j \frac{z_i F}{\kappa RT} D_i C_i = -D_i \frac{dC_i}{dx} + \frac{z_i F}{RT} D_i C_i \frac{d\Phi}{dx} \tag{12}$$

The membrane tends to swell if it has any ionic group ($-SO_3H$) incorporated into it and is immersed in water. This leads to the release of the counter ion (H^+) while the fixed ion (of opposite charge) remains bonded covalently to the membrane structure. If the membrane is placed in an aqueous electrolyte, some salt might enter the membrane. The concentration of co-ions will increase as the concentration of the electrolyte increases. Hence, a concentration gradient is formed when electricity is passed. This is the result of the movement of the counterions through the membrane from one side to another.

Ion transfer in CEM is carried out under conditions of neutral pH in metal salt solutions [28]. Concentrations of all cations (H^+, K^+) at the interface of anode and electrolytes will decrease. This is because all the cations contribute to the charge transfer.

When an ion is near to a membrane, it exhibits flux which is expressed as:

$$j_i = Fz_i N_i = -Fz_i D_i \frac{dC_i}{dx} + \frac{z_i^2 F^2}{RT} D_i C_i \frac{d\Phi}{dx} = -Fz_i D_i \frac{dC_i}{dx} \mid M - t_{im} \kappa_i \frac{d\Phi}{dx} \mid M \tag{13}$$

Flux at the surface of the membrane for a specific ion, usually cation is given by:

$$\frac{jt_+}{F} + \kappa_{LM}[C - C_M] = \frac{jt_+ M}{F} \tag{14}$$

Here, k_{LM} - mass transfer coefficient

When the current density is low, the overall effect would be small. When the current density is large enough, the concentration of counterions becomes zero at the interface, leading to the polarization of the membrane, depicted in Fig (4).

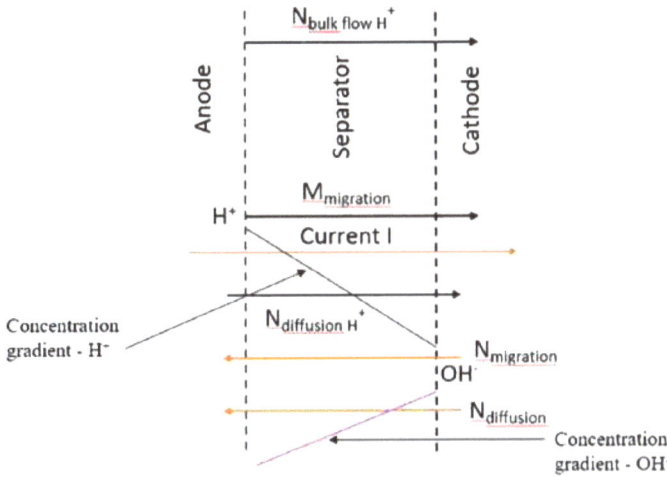

Fig. (4). Concentration profiles at cation-exchange membranes.

Ion and Mass Transfer Processes across Ion-Exchange Membranes in MFCs

Cation Transport

MFC containing CEM will preferentially allow the transport of cations, which leads to the accumulation of protons in the anode. Hence, a wide difference in pH is observed between the anode and cathode. Several problems arise due to this reason [29]. These can be:

- H^+ ions might get accumulated in the anode and the microbial activity gets reduced.
- Carbonate salt might form on the air-cathode due to the higher pH in the cathode caused by the oxidation-reduction reaction. This affects the functioning of the cathode performance.

If the anode is acidic, biofilm development on the membrane surface could be prevented [30, 31]. Migration due to electric field causes transfer of protons through the transference number of a proton, t_p, by:

$$t_p = C_p \lambda_p / \left[\sum i \mid Z_i \mid C_i \lambda_i \right] \tag{15}$$

Here, C_p - concentration of proton, C_i - concentration of ion i, λ_p- molar ionic conductivities of a proton, λ_i - molar ionic conductivities of ion i, pH buffers will largely influence the transfer of ions in MFCs [32].

Anion Transfer

The transfer of OH- ion through the AEM in a basic environment depends on the pH of the cathode. In an air-cathode, bicarbonate and carbonate ions are also present. This is because CO_2 in the air reacts with any OH- ion which is formed during the oxidation-reduction reaction. Anions are transported at much higher concentrations than that of OH- ions when buffers are used in the cathode.

If the pH is quite high, then the hydroxide ion will be transported at a low rate and buffer anions will diffuse through the AEM. The potential of the buffer will become low if the cathode is in batch mode. Hence, to bring the MFC back to the continuous mode, the buffer species have to be renewed [33].

To increase the permeability of the substrate, ion transport has to be increased by using a pH buffer in the anode. This will give a greater permeability to AEM than CEM. Thus, the anions in the anode will diffuse across the AEM and then are transported through migration from cathode to anode [34].

Hence, to enable the transport of protons, the AEM might use a proton carrier and pH buffer in the form of phosphate or carbonate [28, 32]. Phosphate anions show the potential of an effective buffer to balance the pH in MFCs. Eventually, the functioning of the MFC is stimulated [35].

POROUS SEPARATORS

These membranes/separators allow the passage of any ions through it and hence, are not ion-specific. They have the pores all over their structures, which may vary in size. Depending on the size of the pores, these separators can be microporous filtration membranes and ultrafiltration membranes (UFM) [36]. Fabrics, glass fibre, nylon mesh, cellulose filters are generally used as the pore filter materials.

Microporous filtration membranes are mostly used in wastewater treatment as they are much durable and efficient [37]. They allow the passage of charged or neutral species across them depending on the pore size. Larger-sized ions or species will not be successful in passing through the pores. This type of filtration does not completely filter all the species present. Hence, we use the ultrafiltration membranes. In this, pressure is applied to force water through the semi-permeable membrane. This filtration completely gets rid of all the suspended species. These materials are much cheaper than ion-exchange membranes and hence, are used in wastewater treatment.

Many other cheaper membranes like polymer microfilters, nylon, cellulose, and polycarbonate also can act as separators in MFCs. But these have certain disadvantages, the main ones being the permeation of oxygen, and ionic resistance. Adoption of large electrode spacing, and a thick membrane will help in attaining low oxygen permeation. But this will also cause internal resistance and will eventually lower the power density. To eliminate this problem, nanoporous polymer filters usually made of polycarbonate, nylon, polyester, or polyamide, have been proven to be effective in MFCs [38]. From this, it can be concluded that any permeable material can be used for the separator in MFC, provided it allows charge transfer. It should also have the potential to avoid short circuits.

When the pore size of a filter is large, a high flux of oxygen is observed, which tends to deteriorate the functioning of MFCs. This difficulty can be reduced by developing a biofilm on its surface. But this also poses a limitation. The biofilm might eat up the substrate, reduce the surface area, and increase the thickness of the separator. Hence, we can say that the separator is biodegradable.

Glass fibre separators are non-biodegradable and thus, used in lead-acid batteries [39]. This is much more effective in comparison to CEM as it is less permeable to oxygen and is also resistant to biofilm development.

MEMBRANE ELECTRODE ASSEMBLIES

Membrane Electrode Assemblies (MEA) is usually formed as a result of bonding between the electrode and the membrane, to reduce the electrode spacing and the resistance. In an MFC, either the cathode can be bonded to the membrane or the anode can be bonded to the membrane [40]. When the cathode is bonded to the membrane, the anode is located at a position away from the membrane [41]. If the anode is placed on the surface of the membrane, it will come into direct contact with the oxygen that diffuses through the membrane and this is against the anaerobic nature of the anode [42]. The formation of biofilm on the membrane also becomes prevalent which will eventually affect the MFC performance. In the case of bonding, the anode to the membrane also arises certain limitations [43].

One of the combinations can be the bonding of an ion-exchange membrane and a catalyst layer containing an electronic conductor and a binder (Fig. **5**). The electronic conductor is made of carbon and provides a high surface area, and the catalyst is placed on this conductor [44]. The binder is made of an inert polymer powder such as PTFE or PVDF. An ion-exchange material (ionomer) of the same type as the membrane is also used to increase the three-phase reaction zone. Reduction of oxygen takes place in this area, and the electrons and ions are further transported [45].

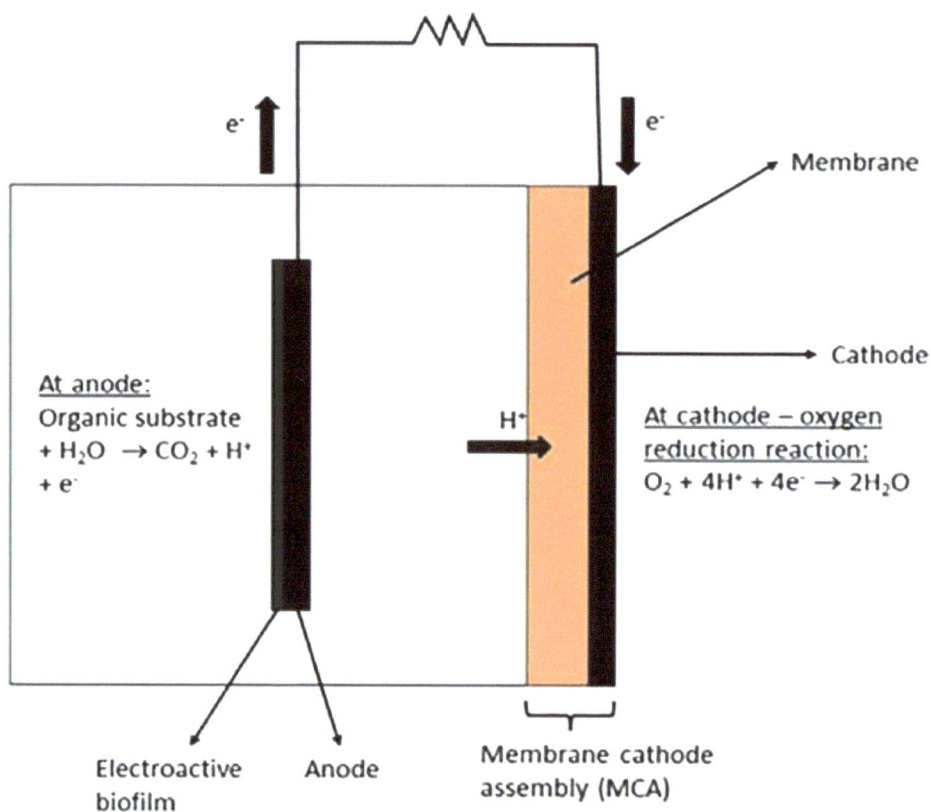

At anode:
Organic substrate
$+ H_2O \rightarrow CO_2 + H^+$
$+ e^-$

At cathode – oxygen
reduction reaction:
$O_2 + 4H^+ + 4e^- \rightarrow 2H_2O$

Membrane

Cathode

Electroactive
biofilm

Anode

Membrane cathode
assembly (MCA)

Fig. (5). Schematics of single chambered microbial fuel cell with membrane cathode assembly.

MEMBRANE FOULING

Membrane fouling is the collection, adsorption or deposition of contaminants on the surface of membranes or cause clogging of membrane pores which can cause a decrease in the filtration capacity of the membrane over time. Other factors like solute removal efficiency, permeate flow and pressure drop across the membrane are also affected due to membrane fouling. Biofouling is the accumulation or attachment of microbes in the raw water, forming a biofilm, on the surface of membrane (Fig. **6**). This contributes majorly to the fouling of RO membranes and can occur at any point during the process of desalination. *Mycobacterium, Flavobacterium, Pseudomonas, Bacillus, Cytophaga* and *Lactobacillus* are some of the common microorganisms capable of forming biofilms on the surface of membranes [46].

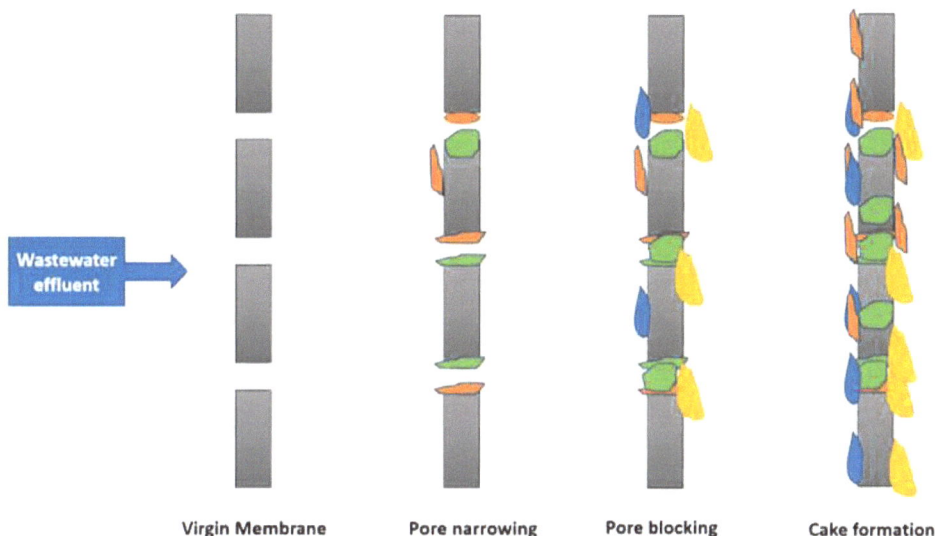

Fig. (6). Cake resistance development on a fouled membrane.

Biofouling degrades the polyamide selective layer through hydrolysis and increases membrane transport resistance. The presence of Extracellular Polymeric Substances (EPSs) promotes the maintenance of biofilms, and are responsible for the process of biofouling. Biofouling causes a variety of negative implications on the process of desalination. In some cases, the water quality reduces drastically due to the entry of the microbes into the water [47]. Another problem posed is the biodegradation of the membrane specifically cellular acetate, due to the production of acids by the biofilm. Recently, many approaches have been implemented to overcome the problem of biofouling. Physical and chemical approaches were developed to prevent attachment of microbes and formation of biofilms and avoid conditions that allow the further growth of microbes. Cleaning in place, pretreatment, spacer design, application of magnetic field and membrane modification are some of the approaches. Pre-treatment options have been employed to reduce the problem of biofouling in RO. Physical treatment techniques like ozonation, ultraviolet radiation and filtration have been commonly used to decelerate the process of biofouling [48].

CONCLUSION

Throughout this paper, we have understood the importance of membranes and their effect on the efficiency of Microbial Fuel Cells (MFCs). A membrane is usually any medium that is used to separate the electrodes and obtain two different compartments. The flow of substances through these compartments gets minimized when they encounter the membrane. The membrane material depends on the pore size, porosity, capacity of ion exchange. Membranes should selectively allow the passage of ionic species through it. Various polymer materials are tested for their potentials to be used in membranes. Ion-exchange membranes are the most important and efficient commercially used membranes. But they too have certain limitations in different factors. CEM is more stable than AEM in electrochemical processes. AEM is unstable in alkaline conditions and hence, other factors are considered while designing a potent AEM which could gain stability in an alkaline environment. The membrane requirements and related disadvantages have to be kept in mind while choosing a material to design a membrane. Every material is tested for its resistance in the cell, current density, conductivity, and permselectivity, before its incorporation into the membrane. This ensures the potency of the membrane as well as increases the MFC performance. We also understood the mechanism of the transfer of ionic species across the membrane. Membrane Electrode Assembly (MEA) is a new approach in which the electrode is bonded to the membrane to reduce the electrode spacing and the resistance. All the new proposals in this concept have the only intention of increasing the efficiency of the MFC. There is an importance in the further development of new materials for separators and membranes in MFCs. Research should be conducted to explore and design various materials which can be able to accomplish an appropriate balance between ion transfer and oxygen permeation. This has been a major problem in most of the membranes. There has been a good improvement in the field of porous separator materials which provides a trustworthy platform for solving this problem. Sulfonated ion-exchange membranes need to be available at low affordable prices. Practical application of MFCs will be sensible only if strong materials enable the modification of membrane arrangements and it can be available at a low cost.

ACKNOWLEDGEMENT

Declared none.

REFERENCES

[1]　Serrano E, Rus G, García-Martínez J. Nanotechnology for sustainable energy. Renew Sustain Energy Rev 2009; 13(9): 2373-84.
[http://dx.doi.org/10.1016/j.rser.2009.06.003]

[2]　Itoshiro R, Yoshida N, Yagi T, Kakihana Y, Higa M. Effect of ion selectivity on current production in sewage microbial fuel cell separators. Membranes (Basel) 2022; 12(2): 183.
[http://dx.doi.org/10.3390/membranes12020183] [PMID: 35207104]

[3]　Sugioka M, Yoshida N, Yamane T, *et al.* Long-term evaluation of an air-cathode microbial fuel cell with an anion exchange membrane in a 226L wastewater treatment reactor. Environ Res 2022; 205: 112416.
[http://dx.doi.org/10.1016/j.envres.2021.112416] [PMID: 34808126]

[4]　Debe M K. Electrocatalyst approaches and challenges for automotive fuel cells. Nat 2012; 486(7401): 43-51.
[http://dx.doi.org/10.1038/nature11115]

[5]　Carrette L, Friedrich KA, Stimming U. Fuel cells - fundamentals and applications. Fuel Cells (Weinh) 2001; 1(1): 5-39.
[http://dx.doi.org/10.1002/1615-6854(200105)1:1<5::AID-FUCE5>3.0.CO;2-G]

[6]　Yang E, Chae KJ, Choi MJ, He Z, Kim IS. Critical review of bioelectrochemical systems integrated with membrane-based technologies for desalination, energy self-sufficiency, and high-efficiency water and wastewater treatment. Desalination 2019; 452: 40-67.
[http://dx.doi.org/10.1016/j.desal.2018.11.007]

[7]　Zhang H, Sun C. Cost-effective iron-based aqueous redox flow batteries for large-scale energy storage application: A review. J Power Sources 2021; 493: 229445.
[http://dx.doi.org/10.1016/j.jpowsour.2020.229445]

[8]　Ramirez-Nava J, Martínez-Castrejón M, García-Mesino RL, *et al.* The implications of membranes used as separators in microbial fuel cells. Membranes (Basel) 2021; 11(10): 738.
[http://dx.doi.org/10.3390/membranes11100738] [PMID: 34677504]

[9]　Leong JX, Daud WRW, Ghasemi M, Liew KB, Ismail M. Ion exchange membranes as separators in microbial fuel cells for bioenergy conversion: A comprehensive review. Renew Sustain Energy Rev 2013; 28: 575-87.
[http://dx.doi.org/10.1016/j.rser.2013.08.052]

[10]　Haile SM. Fuel cell materials and components. Acta Mater 2003; 51(19): 5981-6000.
[http://dx.doi.org/10.1016/j.actamat.2003.08.004]

[11]　Mauritz KA, Moore RB. State of understanding of nafion. Chem Rev 2004; 104(10): 4535-86.
[http://dx.doi.org/10.1021/cr0207123] [PMID: 15669162]

[12]　Wee JH. Applications of proton exchange membrane fuel cell systems. Renew Sustain Energy Rev 2007; 11(8): 1720-38.
[http://dx.doi.org/10.1016/j.rser.2006.01.005]

[13]　Gottesfeld S. Electrocatalysis of oxygen reduction in polymer electrolyte fuel cells: a brief history and a critical examination of present theory and diagnostics. In: Koper MTM, Ed. Fuel cell catalysis. John Wiley & Sons, Inc. 2009; pp. 1-30.
[http://dx.doi.org/10.1002/9780470463772.ch1]

[14]　Zaidi SMJ. Research trends in polymer electrolyte membranes for PEMFC. In: Zaidi SMJ, Matsuura T, Eds. Polymer Membranes for Fuel Cells. Boston, MA: Springer 2009; pp. 7-25.
[http://dx.doi.org/10.1007/978-0-387-73532-0_2]

[15]　Scott K. Handbook of industrial membranes. Elsevier 1995.

[16]　Scott K. Membranes and separators for microbial fuel cells. In: Scott K, Yu EH, Eds. Microbial electrochemical and fuel cells: fundamentals and applications. Woodhead Publishing 2016; pp. 153-78.
[http://dx.doi.org/10.1016/B978-1-78242-375-1.00005-8]

[17]　Scott K. Microbial fuel cells: transformation of wastes into clean energy. In: Gugliuzza A, Basile A, Eds. Membranes for clean and renewable power applications. Woodhead Publishing 2014; pp. 266-300.
[http://dx.doi.org/10.1533/9780857098658.4.266]

[18]　Vielstich W, Lamm A, Gasteiger H. Handbook of Fuel Cells: Fundamentals. Wiley 2003; Vol 3.

[19]　Cleghorn A, Kolde J, Liu W. Catalyst coated composite membranes. In: Vielstich W, Lamm A, Gesteiger HA, Eds. Handbook of Fuel Cells. Chichester: John Wiley 2003; Vol 3.

[20]　Smitha B, Sridhar S, Khan AA. Solid polymer electrolyte membranes for fuel cell applications—a

review. J Membr Sci 2005; 259(1-2): 10-26.
[http://dx.doi.org/10.1016/j.memsci.2005.01.035]

[21]   Mehta V, Cooper JS. Review and analysis of PEM fuel cell design and manufacturing. J Power Sources 2003; 114(1): 32-53.
[http://dx.doi.org/10.1016/S0378-7753(02)00542-6]

[22]   Robertson NJ, Kostalik HA IV, Clark TJ, Mutolo PF, Abruña HD, Coates GW. Tunable high performance cross-linked alkaline anion exchange membranes for fuel cell applications. J Am Chem Soc 2010; 132(10): 3400-4.
[http://dx.doi.org/10.1021/ja908638d] [PMID: 20178312]

[23]   Mamlouk M, Scott K, Horsfall JA, Williams C. The effect of electrode parameters on the performance of anion exchange polymer membrane fuel cells. Int J Hydrogen Energy 2011; 36(12): 7191-8.
[http://dx.doi.org/10.1016/j.ijhydene.2011.03.074]

[24]   Varcoe JR, Slade RCT. Prospects for alkaline anion-exchange membranes in low temperature fuel cells. Fuel Cells (Weinh) 2005; 5(2): 187-200.
[http://dx.doi.org/10.1002/fuce.200400045]

[25]   Merle G, Wessling M, Nijmeijer K. Anion exchange membranes for alkaline fuel cells: A review. J Membr Sci 2011; 377(1-2): 1-35.
[http://dx.doi.org/10.1016/j.memsci.2011.04.043]

[26]   Sata T, Tsujimoto M, Yamaguchi T, Matsusaki K. Change of anion exchange membranes in an aqueous sodium hydroxide solution at high temperature. J Membr Sci 1996; 112(2): 161-70.
[http://dx.doi.org/10.1016/0376-7388(95)00292-8]

[27]   Scott K. Electrochemical reaction engineering. Academic Press 1991.

[28]   Harnisch F, Warmbier R, Schneider R, Schröder U. Modeling the ion transfer and polarization of ion exchange membranes in bioelectrochemical systems. Bioelectrochemistry 2009; 75(2): 136-41.
[http://dx.doi.org/10.1016/j.bioelechem.2009.03.001] [PMID: 19349214]

[29]   Rozendal RA, Hamelers HVM, Buisman CJN. Effects of membrane cation transport on pH and microbial fuel cell performance. Environ Sci Technol 2006; 40(17): 5206-11.
[http://dx.doi.org/10.1021/es060387r] [PMID: 16999090]

[30]   Ghassemi H, McGrath JE, Zawodzinski TA Jr. Multiblock sulfonated–fluorinated poly(arylene ether)s for a proton exchange membrane fuel cell. Polymer (Guildf) 2006; 47(11): 4132-9.
[http://dx.doi.org/10.1016/j.polymer.2006.02.038]

[31]   Ghasemi M, Wan Daud WR, Ismail M, *et al.* Effect of pre-treatment and biofouling of proton exchange membrane on microbial fuel cell performance. Int J Hydrogen Energy 2013; 38(13): 5480-4.
[http://dx.doi.org/10.1016/j.ijhydene.2012.09.148]

[32]   Fan Y, Hu H, Liu H. Sustainable power generation in microbial fuel cells using bicarbonate buffer and proton transfer mechanisms. Environ Sci Technol 2007; 41(23): 8154-8.
[http://dx.doi.org/10.1021/es071739c] [PMID: 18186352]

[33]   Kim JR, Cheng S, Oh SE, Logan BE. Power generation using different cation, anion, and ultrafiltration membranes in microbial fuel cells. Environ Sci Technol 2007; 41(3): 1004-9.
[http://dx.doi.org/10.1021/es062202m] [PMID: 17328216]

[34]   Zuo Y, Cheng S, Logan BE. Ion exchange membrane cathodes for scalable microbial fuel cells. Environ Sci Technol 2008; 42(18): 6967-72.
[http://dx.doi.org/10.1021/es801055r] [PMID: 18853817]

[35]   Cao X, Huang X, Liang P, *et al.* A new method for water desalination using microbial desalination cells. Environ Sci Technol 2009; 43(18): 7148-52.
[http://dx.doi.org/10.1021/es901950j] [PMID: 19806756]

[36]   Logan BE. Microbial fuel cells. Wiley Blackwell 2008.

[http://dx.doi.org/10.1002/9780470258590]

[37] Zuo Y, Cheng S, Call D, Logan BE. Tubular membrane cathodes for scalable power generation in microbial fuel cells. Environ Sci Technol 2007; 41(9): 3347-53.
[http://dx.doi.org/10.1021/es0627601] [PMID: 17539548]

[38] Biffinger JC, Ray R, Little B, Ringeisen BR. Diversifying biological fuel cell designs by use of nanoporous filters. Environ Sci Technol 2007; 41(4): 1444-9.
[http://dx.doi.org/10.1021/es061634u] [PMID: 17593755]

[39] Zhang X, Cheng S, Wang X, Huang X, Logan BE. Separator characteristics for increasing performance of microbial fuel cells. Environ Sci Technol 2009; 43(21): 8456-61.
[http://dx.doi.org/10.1021/es901631p] [PMID: 19924984]

[40] Yu EH, Burkitt R, Wang X, Scott K. Application of anion exchange ionomer for oxygen reduction catalysts in microbial fuel cells. Electrochem Commun 2012; 21(1): 30-5.
[http://dx.doi.org/10.1016/j.elecom.2012.05.011]

[41] Khilari S, Pandit S, Ghangrekar MM, Pradhan D, Das D. Graphene oxide-impregnated pva–sta composite polymer electrolyte membrane separator for power generation in a single-chambered microbial fuel cell. Ind Eng Chem Res 2013; 52(33): 11597-606.
[http://dx.doi.org/10.1021/ie4016045]

[42] Liu H, Logan BE. Electricity generation using an air-cathode single chamber microbial fuel cell in the presence and absence of a proton exchange membrane. Environ Sci Technol 2004; 38(14): 4040-6.
[http://dx.doi.org/10.1021/es0499344] [PMID: 15298217]

[43] Zhuang L, Zhou S, Wang Y, Liu C, Geng S. Membrane-less cloth cathode assembly (CCA) for scalable microbial fuel cells. Biosens Bioelectron 2009; 24(12): 3652-6.
[http://dx.doi.org/10.1016/j.bios.2009.05.032] [PMID: 19556120]

[44] Pandit S, Khilari S, Bera K, Pradhan D, Das D. Application of PVA–PDDA polymer electrolyte composite anion exchange membrane separator for improved bioelectricity production in a single chambered microbial fuel cell. Chem Eng J 2014; 257: 138-47.
[http://dx.doi.org/10.1016/j.cej.2014.06.077]

[45] HaoYu E, Cheng S, Scott K, Logan B. Microbial fuel cell performance with non-Pt cathode catalysts. J Power Sources 2007; 171(2): 275-81.
[http://dx.doi.org/10.1016/j.jpowsour.2007.07.010]

[46] Jadhav DA, Pandit S, Sonawane JM, Gupta PK, Prasad R, Chendake AD. Effect of membrane biofouling on the performance of microbial electrochemical cells and mitigation strategies. Bioresour Technol Rep 2021; 15(9): 100822.
[http://dx.doi.org/10.1016/j.biteb.2021.100822]

[47] Ishizaki S, Papry RI, Miyake H, Narita Y, Okabe S. Membrane fouling potentials of an exoelectrogenic fouling-causing bacterium cultured with different external electron acceptors. Front Microbiol 2019; 9: 3284.
[http://dx.doi.org/10.3389/fmicb.2018.03284]

[48] Ghasemi M, Wan Daud WR, Ismail M, *et al.* Effect of pre-treatment and biofouling of proton exchange membrane on microbial fuel cell performance. Int J Hydrogen Energy 2013; 38(13): 5480-4.
[http://dx.doi.org/10.1016/j.ijhydene.2012.09.148]

CHAPTER 6

Polymer Composites for Sensor Applications

Arti Rushi[1], **Kunal Datta**[2] and **Bhagwan Ghanshamji Toksha**[1, *]

[1] *Maharashtra Institute of Technology, Aurangabad, India*

[2] *Deen Dayal Upadhayay KAUSHAL Kendra Dr. Babasaheb Ambedkar Marathwada University, Aurangabad (MS), India*

Abstract: Polymers play a major role in sensor research nowadays. Specifically, when the electrical modality of sensing is concentrated then conducting polymers is found to be highly useful. They have been explored for the development of sensors to cope with advanced modern-day requirements. There is a huge demand for sensors in detecting and assessing environmental dynamics, harmful working conditions, food poisoning, and water contaminations, and diagnostic purposes. The recent pandemic, the COVID-19 outburst all over the world, ascertained the urgency of research in the direction of designing and developing biosensors enabling distinction among the diseases and enabling medical professionals to take faster clinical decisions. The conventional approaches in environment pollutant detection techniques have no universally accepted code of conduct. Moreover, there are various experimental drawbacks of poor calibration, tedious sample preparation, blank determination, and lengthy time-consuming procedure. The composites involving conducting polymers and CNTs bring in unique multifunctional features. The motive of the present work is to review various latest developments in conducting polymer composite-based sensors.

Keywords: Composites, Multi-walled CNT, Polymers, Single-walled CNT, Sensors.

INTRODUCTION

Functional polymer composites are a class of material displaying rapid, reversible, repetitive, and measurable changes in response to any detectable change in the area of interest when employed in detection applications [1]. The characteristics of this class of materials enable the sensor-based devices to respond to an external chemical/physical stimulus. The modern era of bio and chemical sensors has enhanced the facilities significantly in the field of medical diagnosis and environmental probing [2-5]. The critical parameters which decide the performance of bio/chemical sensors are sensitivity to the desired stimuli at the

---

* **Corresponding author Bhagwan Ghanshamji Toksha:** Maharashtra Institute of Technology, Aurangabad, India;
E-mail: bhagwantoksha@gmail.com

Subhendu Bhandari, Prashant Gupta and Ayan Dey (Eds.)

micro-scale selectively, biocompatibility, and quicker response time [6-8]. The aspects of pollution include soil, air and water pollution which may or may not detected visually and by taste. Air pollution target ammonia, CO, NO_2, H_2S and many other gases, while the soil and water pollution mainly target heavy metal ions such as chromium (Cr), cadmium (Cd), mercury (Hg), lead (Pb) and arsenic (As). Other than damaging the respiratory systems, these pollutants are also neurotoxic producing a series of toxicological reactions. These are also responsible for medical complications such as anaemia, seizures, coma or even fatality. The market based device requirements such as cost effectiveness, robustness, miniaturized size and workability in extreme conditions are also desirable to the functioning and eventual performances [9-12]. In the unavailability of all the required parameters for effective sensing applications, there is a possible solution of using a composite or blend of two or more materials. The choice of sensing materials for the monitoring device and the involved active microelectronics becomes the next critical step of the design of sensor. The class of conducting polymers materials is explored in sensor technologies owing to their versatility, ease in production, higher surface area, low cost, high sensitivity when exposed to a variety of target pollutants and simple signal detection in terms of change of electrical and optical responses [13-17]. The carbon allotropes have proven their high potential in sensing applications due to their outstanding properties, especially in nanoscale. The structure of conjugated conducting polymers is such that single and double bonds occur alternatively in their polymer chain, thus forming the delocalized electrons which act as charge carriers. The low conductivity of polymers under normal conditions can be improved by composite effect with an oxidizing or reducing agent to several folds in magnitude [18]. The composite recipes also lead to better sensing performance. Such an improvement could be related to highly porous microstructure and good antifouling activity achieved in composite phases [19]. The inclusion of carbon allotropes in composites with conducting polymer could lead to formation of conducting paths, high active surface area and unique 3D microstructure contributing towards better sensing performance [20-23]. The developments in the field of 3D printing of sensors having conducting polymers and CNT composite as building block brings newer possibilities of sensing functions and expands the application boundaries. The possibilities involve the fabrication of highly customizable sensors capable of detecting microbial activity [24]. The other possibility with 3D print sensors to respond and record magnetic and electric fields, heat, light, pH and humidity. The research carried out in this direction produced promising results exhibiting 3D printed conducting polymers and CNT composite better conductive and mechanical properties and a better performance as compared to other formulations. The sensor functionalities were also achieved with commercially available desktop 3D printer leading to achieving low-cost functional sensor devices [25]. The blend of carbon nanotubes and graphene nanoplatelet in the thermoplastic polyurethane (TPU) composites were reported to result in high-performance flexible strain sensors synthesized *via* Fused Filament Fabrication (FFF) 3D printing. The blend carbon nanotubes and graphene nanoplatelet were reported to produce synergetic effect demonstrating higher sensitivity, better stability, and higher accuracy [26].

POLYMER COMPOSITE

Polymers are the materials which are formed by repetition of smaller chemical units which results into long chain molecules. Polymers has been found naturally, most interestingly; they are the integral part of living organisms. Polymers are found in living species in the form of proteins, cellulose, and nucleic acids. Naturally occurring rubber, various resins are some another examples of polymers. Along with this, synthetic polymers such as polyethylene, polypropylene, polystyrene are also getting the high attraction from entire human race. It has been observed that, polymers developed at early stage are insulating in nature. The conducting properties of the polymers were discovered by A. G. MacDiarmid, Professor A. J. Heeger and Professor H. Shirakawa. For this discovery they are awarded with the Nobel Prize in chemistry in the year 2000. Some of the well-known conducting polymer structures are provided below (Fig. **1**):

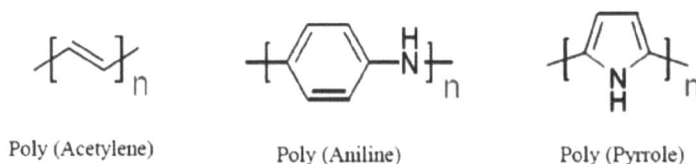

| Poly (Acetylene) | Poly (Aniline) | Poly (Pyrrole) |

Fig. (1). Conducting polymer structures.

Modulation in the conducting properties of the conducting polymers provides an additional benefit for making their use in various electronics applications. Also, synthesizing the nanostructures of conducting polymers and making use of them are fascinating factors for the researchers. As far as sensor applications of conducting polymers are concerned, the nanostructures of conducting polymers are adopted in high extent [27]. Synthesis of composites is another approach that has been adopted to prepare polymer-based materials in the sensing field [28, 29]. It has been seen that polymers can be combined with the materials such as graphene [30], carbon nanotube [31] *etc.*, or the composites can also be formed by combining two different polymers [32]. Most of the sensors based on polymer composite, are fabricated with conducting polymers and carbon nanotubes.

Considering this point, a detailed discussion on carbon nanotubes and their composites with polymers has been given in the successive sections.

CARBON NANOTUBES IN SENSING FIELD

SumioIijima in 1991 [33] discovered carbon nanotubes, and from then, a new research era has started which contributed positively in the wellbeing of the entire human race. Carbon nanotubes (CNTs) are the allotropes of carbon which are the seamless rolling of graphene sheet(s). Depending upon the number of layers rolled while forming the CNTs, three types are generated as: single walled carbon nanotubes (SWCNTs), multiwalled carbon nanotubes (MWCNTs) and double walled carbon nanotubes (DWCNTs). The carbon nanotubes discovered by Ijima in 1991, were the MWCNTs, and he himself with his co-workers [34] in 1993 discovered the SWCNTs. In carbon nanotubes, the atoms of carbon are arranged in a hexagonal manner in such a way that it forms a tubular structure whose diameter is approximately 1nm and its length lies in between 1 μm to 100 μm. The dimensions and properties of CNTs are proportional to the synthesis parameters.

For the synthesis of CNTs, basically three methods are adopted: i) arc discharge method [35], ii) laser ablation method [36, 37], and iii) chemical vapour deposition (CVD) [38]. The fabricated CNTs need to undergo numerous purification processes. The methodology of the purification procedure depends on the synthesis techniques. Carbon nanotubes synthesized by arc discharge, laser ablation or CVD method contains different scales of impurities. Impurities are of various kinds like carbon material's soot, metal particles *etc.* The purification techniques such as gas phase purification [39] and liquid phase purification [40] can be employed for CNTs synthesized by the arc discharge method. The techniques such as microscale and macroscale purification can be used for the nanotubes grown in laser ablation method [41]. For the carbon nanotubes grown with CVD, purification procedures such as sonication, oxidation and acid washing [42] are applied. Although, purification step is of high importance before making use of synthesized CNTs, this may also damage the sidewall surfaces in a small amount. The purity status of the CNTs can be determined with the spectroscopic tool such as Raman spectroscopy [43]. The emphasis of the following discussion will be on the CNTs' potential for gas sensing, rather than exploring all of their remarkable features. Mainly, the focus will be given on sensing properties of the SWNT-based sensor in chemically effective field effect transistor (chemFET) / chemiresistor-based modality. Ease of fabrication and easy mode of operation are the reasons. The discussion provided below is related to the chemFET and CNT-based chemiresistor. Josef Christ *et al.* reported an innovative application of wearable electronics. In this work, a multi-layer piezoresistive sensor was achieved with a thermoplastic polyurethane and MWCNT [44]. The platform was produced by an extrusion process followed by 3D printing. A direction-independent repeatable piezoresistive response for measuring finger flexure is presented in Fig (**2**).

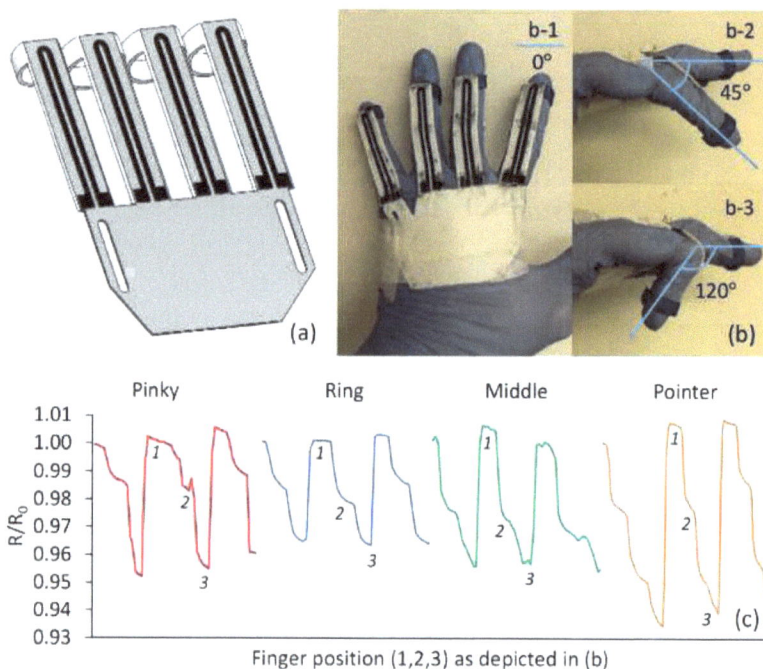

Fig. (2). Polymer + MWCNT composite piezoresistive responses for the measuring finger flexure sensor, prototype and cyclic presentation graph [44].

According to the structural calculations, a SWNT should have a minimum diameter of 0.4 nm and a maximum diameter of 3.0 nm in order to preserve its tubular structure as it forms from a graphite sheet [45, 46]. Unless it is sustained by another force or surrounded by the nearby force, as in MWNTs, a bigger diameter SWNT will collapse. Due to electron confinement and a reduction in the phase space for scattering caused by such a drastically reduced cross section, there is a low chance of scattering and strong mobility for charge carriers. Such characteristics in semiconductor CNT transistors may result in an ON current of 1 mA/m. These facts point to a faster charge transduction in CNTs, and when used as a gas sensor, these properties drastically shorten the reaction time for CNTs-based sensors. At the same time, CNTs can survive challenging operating circumstances because of their superior mechanical/thermal characteristics and chemical stability [47]. However, the most favourable feature of the 1-D nanostructure of CNTs is its extremely high aspect ratio (surface-to-volume ratio), which results in the presence of all atoms on the tube's surface and has a diameter of just a few nanometers and length up to 100 microns. Since any local charge can significantly alter the carrier concentration along the 1-D wire axis, such molecular wires [48] with excellent quantum confinement are particularly sensitive to the local environment. As a result, CNTs emerge as one of the most effective nanostructured sensing elements. Like this, the circular curvature in CNTs gives σ-π rehybridization in which three σ bonds are somewhat out of plane. The π orbital is now asymmetrically distributed both within and outside the cylindrical wall of the nanotube as correction.

Similarly, the circular curvature in CNTs results in σ-π rehybridization in which three σ bonds are slightly out of the plane. Now, for compensation, the π-orbital is asymmetrically distributed inside and outside the cylindrical wall of nanotube. Because of the deformed electron clouds, a rich π- -electron conjugation forms outside the tube, making CNTs extremely electrochemically active. NO_2, NH_3, O_2, and other chemicals that may donate or take electrons from surfaces of CNTs will do so, modifying the overall conduction behavior [49]. The following is an example of an electrochemical reaction that occurs when CNT and an electron-donating or -withdrawing analyte interact:

$$CNT + Gas \longrightarrow CNT^{\delta-}Gas^{\delta+} \text{ or } CNT^{\delta+}Gas^{\delta-}$$

$$CNT + Gas \rightarrow CNT^{\delta-}Gas^{\delta+} + CNT^{\delta+}Gas^{\delta-}$$

where δ denotes the number of charges transferred during interaction.

The above electron donation/withdrawal from the nanotube's surface or inside causes well-mechanized effects. The overall sensing phenomena is shaped by enhanced molecule adsorption at room temperature and subsequent charge transfer. Gas molecules may be adsorbed in tube bundle interstices, grooves above gaps between adjacent tubes, nanopores inside tubes, and surfaces of individual tubes.

In terms of the carbon nanotubes' ability to detect pollutants, Kong *et al.* were the first to describe the simultaneous detection of NO_2 and NH_3 using a SWNT FET [50]. This particular endeavour was a logical progression of a finding from Zhou *et. al.* [51] where the FET characteristics of SWNTs integrated in a back-gated configuration were described by the authors. Kong *et. al.* [50] had noticed three orders of change in SWNT conductance within one or two minutes to even 2 ppm of NO_2 and 0.1% of NH_3 concentration. Someya *et al.* [52] reported using a similar modality of sensing backbone to detect saturated vapours of methanol, ethanol, 1-propanol, 2-propanol, and tertiary-butanol with response times between

5 and 15 seconds. In order to detect dimethyl methylphosphonate, Novak *et al.* [53] showed a CHEMFET based on a SWNTs network (DMMP). In this instance, a sub-ppm level reading was reported. The study's most important component was the sensor's quick recovery after using a positive gate bias. The results of Chang *et al.* later helped to further establish this recovery approach [54]. Valentini *et al.* constructed a chemiresistor-based sensor using CNT mats [55]. that had a 10 ppb NO_2 detection limit. The sensor was examined, nonetheless, for a high operating temperature. Li *et al.* created an easy-to-make SWNT-based chemiresistor by simply drop-casting well-dispersed SWNTs onto interdigitated Au electrodes that had already been created [56]. The sensors were successfully used for nitrotoluene and NO_2 at ppb level monitoring. Except when they were exposed to UV radiation for quick recovery, the sensors recovered slowly. Suehiro *et al.* were the ones who initially used dielectrophoresis [57] and they have developed an ammonia sensor based on MWNT network. Li *et al.* [58] reported direct CVD synthesis of SWNTs on Si/SiO_2 substrate with integrated Pt and Ti electrodes. The sensors showed NO_2 and NH_3 concentration detection down to 1 and 125 ppb, respectively. Goldoni *et al.* reported the impact of contaminants created during the synthesis and purification of SWNTs on their sensing capabilities [59]. A CNT film-based H_2 sensor with a detection limit of 10 ppm at room temperature was made, according to Pierton *et al.* [60]. According to Matranga and Bockrath, CO can be detected by forming a hydrogen bond with the hydroxyl groups on CNT [61]. SWNTs/MWNTs have also been researched for various approaches including chemicapacitors in addition to the more commonly reported chemiresistor or CHEMFET modality sensors [62], surface acoustic wave (SAW) [63], microwave resonant sensor, *etc* [64]. A thorough discussion on CNT-based sensors has been provided by Zhang *et. al* [65], Schroeder *et al.* [66], Maryam *et. al.* [67] and Star *et al.* [68].

The explanation of pure SWNTs and MWNTs-based sensors as mentioned above makes it evident that there is one major issue with useable CNT sensors, namely, the extreme restriction in the potential transduction mechanism. A wide range of efforts have been made so far in order to increase the adaptability of CNT-based sensors and make them capable of sensing a wider range of analytes, including (but not limited to) (i) creating defect sites on CNTs surface [69], (ii) doping in CNTs [70], (iii) functionalization of CNT side wall with target specific entities that contain porphyrins [71], metal oxides [72], *etc.* and (iv) formation of composites [73] *etc.* Since the analyte is prevented from coming into direct contact with the CNT surface in this configuration, functionalizing the CNT sidewall or creating a composite also has the added benefit of boosting sensor recovery characteristics. It is true that the CNTs' honeycomb shape and the high binding energy of gas analytes cause analytes to slowly desorb, which reflects poor recovery behavior in pristine CNT-based sensors. As a result, composite

CNT structures have generated a lot of research attention in efforts to solve selectivity and recovery issues with CNT-based sensors. In general, CNT composites with a conducting polymer have attracted a lot of interest lately. Below is a full description of CNT-polymer composite-based sensors.

CNT- POLYMER COMPOSITE GAS SENSORS

It is clear from our previous discussions that carbon nanotubes, with their extremely high aspect ratio, higher electrochemical activity, and exceptional electrical conduction characteristics at room temperature, are the most effective gas sensing entities. The biggest obstacles to commercial sensing solutions utilising CNTs, however, are limited number of transduction mechanisms available and the high binding affinity of CNTs. As discussed earlier, conducting polymers (CPs) can be most interesting solution to overcome the problems associated with CNT. The functional composites of carbon nanotubes and conducting polymer have gained high research interest. A recent study on the phthalocyanines/SWCNTs-COOH hybrid substitute with peripheral alkoxy reported effective, reversible and fast response in toxic gas sensing at room temperature [70, 74-76].

Conducting polymers and their nanostructures are found to be attractive solutions in the fabrication of the electronic sensing devices. Nanostructured forms of conducting polymers due to their modulating electrical conducting properties, flexible nature, chemical and electrochemical activeness, and easy synthesis approach gained high attention of the research community. Among these mentioned properties, ease in modulation of electrical properties of conducting polymer by chemical route, make them promising sensing material. With nanostructured forms of conducting polymers, detection of gaseous components, biological elements and chemical components becomes very easy. The detailed discussion on pollutant detection properties of conducting polymer composites was done by Shirsat *et al.* [77]. Detection of NH_3, NO_2, H_2S, and CO was explained in detailed manner. Basically, following routes are adopted to form the composites of carbon by functionalization with different functional groups on the surface of carbon nanotubes:

i. Covalent functionalization of sidewalls of carbon nanotubes.
ii. Non-covalent exohedral functionalization of carbon nanotubes by conducting polymer nanostructures.

Both covalent and non-covalent methods can be used to create composite materials out of CNTs and CPs. Functional composites of CNTs-CP are formed through non-covalently generated structures. Since the chemical treatment required to make a CNT-CP functional composite *via* the covalent method is frequently harsh and interferes with the surface activity of the CNTs, this method is less desirable [78]. Contrarily, CNT-CP functional composite generated by a non-covalent method is extremely favourable for the reasons listed below:

i. Adsorptive and wrapping forces, such as Van der Waals and -stacking interaction, are commonly used in supramolecular techniques for non-covalent functionalization because they increase the possibility of wrapping up the entire SWNT surface while preserving their attractive surface activity [79].
ii. Since CNTs and CPs have comparable structures, aromatic polymer rings have a propensity to gradually align their ring-planes parallel to the SWNT surfaces during dynamic interactions [80]. As a result, the CNT-CP interface forms a highly effective -conjugation that provides a route for charge carriers along the delocalized backbone.
iii. A wide variety of polymers are available - thanks to the synthetic diversity of coating polymers, and can be applied to the surface of SWNTs using various methods. Electrochemical approaches, in particular, provide more control over the production and thickness of the polymeric layer. The fields of a biosensor and biofuel cell applications are benefited greatly from the use of functionalized nanocomposites. The low functionality SWNTs were dissolved in water using a variety of proteins, including cytochrome c, horseradish peroxidase (HRP), and bovine serum albumin (BSA). The sensor developed in this way performed better in detecting hydrogen peroxide on an electrode modified with polypyrrole/SWCNT-HRP nanocomposites [74, 81, 82]. The detailed procedure for the fabrication of SWNT – CP composite-based chemiresistors has been explained by Ghosh *et al.* [83]. They have reported the composite of poly N-methyl pyrrole nanowire with carbon nanotubes. After the formation of conducting polymer and SWNT composite, various characterization tools can be applied to understand the properties of the composite. Raman Spectroscopy is one of the tools with which physical properties of the composite can be determined. Interfacial properties, mechanical deformations caused in CNTs due to composite formation, phase transitions of the polymer could be determined with Raman spectroscopy [84]. Explanation of conducting properties of CNT-CP composite was given by James T. Wescott [85]. A systematic review on interfacial characteristic of CNT-polymer nanocomposite was reviewed by Chen *et al.* [86]. Table **1** gives an overview of the sensing characteristics of the CNT-CP composites:

Table 1. Gas sensing performance of CNT-CP composite.

Materials	Modality of Sensing	Analyte Detected	Reference
SWNT-Poly N-methyl pyrrole nanowire	Chemiresistive	Ammonia	[83]
SWCNTs-copolymer containing amidine pendant groups	Chemiresistive	CO_2	[87]
Polystyrene-block-poly(tert-butyl methacrylate)/multiwall carbon nanotube	Chemiresistive	volatile organic compounds (VOCs)	[88]
PANI-SWNTs	ChemFET	NH_3, CO	[89]
poly(3,4-ethylenedioxythiophene) - carbon nanotube nanocomposite	Amperometric	nitrobenzene	[90]
CNT-PANI	Chemiresistive	Hydrogen	[91]
5-amino-1,3,4-thiadiazole-2-thiol (ATT) -acid functionalized multiwalled carbon nanotubes (FMWCNTs).	Amperometric	Nitrite	[92]
multiwall carbon nanotubes/conducting polymer	Chemiresistive	NH_3	[93]

From the above table, it is evident that chemiresistive modality of sensing is the most efficient for CNT-CP composite. The selectivity of the CNT-CP composite has been observed towards the gaseous analytes such as ammonia, carbon monoxide, carbon dioxide, volatile organic compounds (VOCs), *etc.* In a polymer-based CNT nano-composite, the structural change brought in CNT could be critical for gas sensing applications. This modification is achieved with an aim to reduce electron work function and resultant composite conductivity [94]. The morphological modifications revealed through SEM and TEM images are presented in Fig (**3**).

CNT-POLYMER COMPOSITE-BASED SENSORS

Other than gas sensing capabilities, polymer composites have been employed for the detection of some other entities which may include some biological components or some physical components. A bolometric sensor has been developed by Aliev *et al* [95]. Authors have fabricated uncooled bolometric sensor based on the SWNTs polymer composite with enhanced sensitivity. Fabrication of flexible, attachable electrochemical DNA sensors was done by Lee *et al.* [96]. They have used carbon-nanotube-based electrodes which have been prepared by simple all-solution processing. The polymer used in making the composite was polydimethylsiloxane (PDMS). Authors concluded that the CNT-polymer composite was found to be user-friendly and easily attached DNA sensors could be prepared at low cost. Chemical sensors with composites of conducting polymers with graphene and carbon nanotube were prepared by

Salavagione *et al.* Authors have done quantitative and qualitative analysis in diversified fields of application such as biosensing (DNA, enzymes, proteins, antigens and metabolites), as well as chemical and gas sensing using electrochemical and optical detection methods [97]. P. Pissis *et al* [98]. investigated on strain and damage sensing with polymer composites. For creation of the sensors, they used PEEK (polyether ether ketone) reinforced with CFs (carbon fibers), SBR (styrene butadiene rubber) filled with CB, and PP (polypropylene) filled with unfunctionalized MWCNTs. Detailed discussion on strain and pressure sensing capabilities of polymer composites has been provided by Kanoun [99]. In that review paper, authors provided recent advancements in polymer/CNT nanocomposites which could be used in the fabrication of pressure and strain sensors. They mainly put their thoughts on sensing principles involved and impact of fabrication procedures on the sensing mechanism of the sensor.

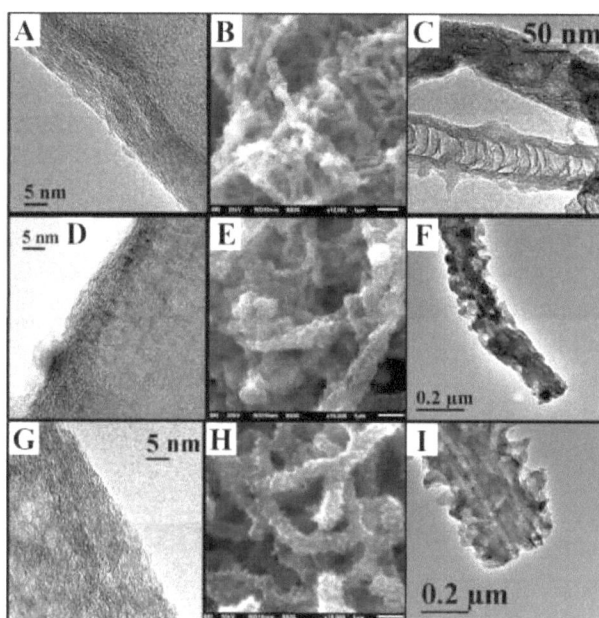

Fig. (3). Morphological modificationsin CNT + Polymer nano-composite [94].

CONCLUSION

The rush of growth has led towards the development of industries and factories. Though the growth is inevitable, the accompanying pollution caused by the output from industries and factories have created a serious challenge towards the pollution of environmental resources. The scientific community has picked up this challenge of monitoring the levels of such pollutants to counter high toxicity and possible hazards to human health. Conducting polymer nanocomposites with carbon allotropes have been highly acclaimed by researchers due to their diverse synthetic methods, and high adsorption capacity in sensing and monitoring contaminants. The particular usefulness of various polymers and underlying detection mechanisms were found to vary for detection of various hazardous environmental pollutants. The polyaniline composites demonstrated better results in ammonia detection while polypyrrole, poly (3,4-thylenedioxythiophene) were useful for CO detection. The sensors based on polythiophene and CNT composites were effective in NO_2 and H_2S monitoring. The unique physicochemical and electrical properties of carbon nanotube/ polymer composite enable these materials to contribute towards monitoring and controlling the environmental pollution along with high sensitivity and precise selective detection. The polymer composites have contributed towards the improvement of pristine conducting polymer or CNT downside properties in regard to efficiency and stability. Apart from gas sensing applications, polymer composites are also beneficial in stress sensing, pressure sensing, and biomaterials sensing.

ACKNOWLEDGEMENT

Declared none.

REFERENCES

[1]   R. Brighenti, Y. Li, and F. J. Vernerey, Smart polymers for advanced applications: a mechanical perspective review, Front. Mater., 7, 2020, 30, 2022. [Online]. Available from: www.frontiersin.org/article/
[http://dx.doi.org/10.3389/fmats.2020.00196]

[2]   Helfmann J, Netz UJ. Sensors in diagnostics and monitoring. Photonics Lasers Med 2015; 4(2): 107-9.
[http://dx.doi.org/10.1515/plm-2015-0012]

[3]   Spook SM, Koolhaas W, Bültmann U, Brouwer S. Implementing sensor technology applications for workplace health promotion: a needs assessment among workers with physically demanding work. BMC Public Health 2019; 19(1): 1100.
[http://dx.doi.org/10.1186/s12889-019-7364-2] [PMID: 31412839]

[4]   Aguilar AJ, de la Hoz-Torres ML, Martínez-Aires MD, Ruiz DP. Monitoring and assessment of indoor environmental conditions after the implementation of covid-19-based ventilation strategies in an educational building in southern spain. Sensors (Basel) 2021; 21(21): 7223.
[http://dx.doi.org/10.3390/s21217223] [PMID: 34770530]

[5]   Martinez Paz EF, Tobias M , Escobar E *et al.* Wireless sensors for measuring drinking water quality in building plumbing: deployments and insights from continuous and intermittent water supply systems. ACS EST Eng 2021; 423-33.
[http://dx.doi.org/10.1021/acsestengg.1c00259]

[6]   Gao Y, Wang Y, Dai Y, *et al.* Amylopectin based hydrogel strain sensor with good biocompatibility, high toughness and stable anti-swelling in multiple liquid media. Eur Polym J 2022; 164: 110981.
[http://dx.doi.org/10.1016/j.eurpolymj.2021.110981]

[7]   Gauns Dessai PP, Singh AK, Verenkar VMS. Mn doped Ni-Zn ferrite thick film as a highly selective and sensitive gas sensor for Cl2 gas with quick response and recovery time. Mater Res Bull 2022; 149: 111699.
[http://dx.doi.org/10.1016/j.materresbull.2021.111699]

[8]   Ghazi M, Janfaza S, Tahmooressi H, Tasnim N, Hoorfar M. Selective detection of VOCs using microfluidic gas sensor with embedded cylindrical microfeatures coated with graphene oxide. J Hazard Mater 2022; 424(Pt C): 127566.
[http://dx.doi.org/10.1016/j.jhazmat.2021.127566] [PMID: 34736204]

[9]   Berdinsky AS, Fink D, Petrov AV, *et al.* Formation and conductive properties of miniaturized fullerite sensors. Proc MRS 2001; 705(1): Y4.7.
[http://dx.doi.org/10.1557/PROC-705-Y4.7]

[10]   Lai J, Vygranenko Y, Heiler G, *et al.* Noise performance of high fill factor pixel architectures for robust large-area image sensors using amorphous silicon technology. Proc MRS 2007; 989(1): 0989-A14-05.
[http://dx.doi.org/10.1557/PROC-0989-A14-05]

[11]   Haines T, Bowles KA. Cost-effectiveness of using a motion-sensor biofeedback treatment approach for the management of sub-acute or chronic low back pain: economic evaluation alongside a randomised trial. BMC Musculoskelet Disord 2017; 18(1): 18.
[http://dx.doi.org/10.1186/s12891-016-1371-6] [PMID: 28095832]

[12]   Lahokallio S, Hoikkanen M, Marttila T, Vuorinen J, Kiilunen J, Frisk L. Performance of a polymer-based sensor package at extreme temperature. J Electron Mater 2016; 45(2): 1184-200.
[http://dx.doi.org/10.1007/s11664-015-4198-2]

[13]   Swager TM. 50th Anniversary perspective: conducting/semiconducting conjugated polymers. a personal perspective on the past and the future. Macromolecules 2017; 50(13): 4867-86.
[http://dx.doi.org/10.1021/acs.macromol.7b00582]

[14]   Dakshayini BS, Reddy KR, Mishra A, *et al.* Role of conducting polymer and metal oxide-based hybrids for applications in ampereometric sensors and biosensors. Microchem J 2019; 147: 7-24.
[http://dx.doi.org/10.1016/j.microc.2019.02.061]

[15]   Bhadra S, Khastgir D, Singha NK, Lee JH. Progress in preparation, processing and applications of polyaniline. Prog Polym Sci 2009; 34(8): 783-810.
[http://dx.doi.org/10.1016/j.progpolymsci.2009.04.003]

[16]   Barboza BH, Gomes OP, Batagin-Neto A. Polythiophene derivatives as chemical sensors: a DFT study on the influence of side groups. J Mol Model 2021; 27(1): 17.
[http://dx.doi.org/10.1007/s00894-020-04632-w] [PMID: 33409576]

[17]   Xu T, Dai H, Jin Y. Electrochemical sensing of lead(II) by differential pulse voltammetry using conductive polypyrrole nanoparticles. Mikrochim Acta 2020; 187(1): 23.
[http://dx.doi.org/10.1007/s00604-019-4027-z] [PMID: 31807912]

[18]   Zhang Z, Liao M, Lou H, Hu Y, Sun X, Peng H. Conjugated Polymers for Flexible Energy Harvesting and Storage. Adv Mater 2018; 30(13): 1704261.
[http://dx.doi.org/10.1002/adma.201704261] [PMID: 29399890]

[19]   Wang W, Cui M, Song Z, Luo X. An antifouling electrochemical immunosensor for carcinoembryonic antigen based on hyaluronic acid doped conducting polymer PEDOT. RSC Advances 2016; 6(91): 88411-6.
[http://dx.doi.org/10.1039/C6RA19169J]

[20]   Liu B, Zhang X, Tian D, *et al. In situ* growth of oriented polyaniline nanorod arrays on the graphite flake for high-performance supercapacitors. ACS Omega 2020; 5(50): 32395-402.
[http://dx.doi.org/10.1021/acsomega.0c04212] [PMID: 33376876]

[21]   Amatatongchai M, Sroysee W, Sodkrathok P, Kesangam N, Chairam S, Jarujamrus P. Novel three-Dimensional molecularly imprinted polymer-coated carbon nanotubes (3D-CNTs@MIP) for selective detection of profenofos in food. Anal Chim Acta 2019; 1076: 64-72.
[http://dx.doi.org/10.1016/j.aca.2019.04.075] [PMID: 31203965]

[22]   Park S, Shou W, Makatura L, Matusik W, Fu KK. 3D printing of polymer composites: Materials,

processes, and applications. Matter 2022; 5(1): 43-76.
[http://dx.doi.org/10.1016/j.matt.2021.10.018]

[23]   Abshirini M, Charara M, Liu Y, Saha M, Altan MC. 3D printing of highly stretchable strain sensors based on carbon nanotube nanocomposites. Adv Eng Mater 2018; 20(10): 1800425.
[http://dx.doi.org/10.1002/adem.201800425]

[24]   Khosravani MR, Reinicke T. 3D-printed sensors: Current progress and future challenges. Sens Actuators A Phys 2020; 305: 111916.
[http://dx.doi.org/10.1016/j.sna.2020.111916]

[25]   Gnanasekaran K, Heijmans T, van Bennekom S, *et al.* 3D printing of CNT- and graphene-based conductive polymer nanocomposites by fused deposition modeling. Appl Mater Today 2017; 9: 21-8.
[http://dx.doi.org/10.1016/j.apmt.2017.04.003]

[26]   Xiang D, Zhang X, Han Z, *et al.* 3D printed high-performance flexible strain sensors based on carbon nanotube and graphene nanoplatelet filled polymer composites. J Mater Sci 2020; 55(33): 15769-86.
[http://dx.doi.org/10.1007/s10853-020-05137-w]

[27]   Yoon H. Current trends in sensors based on conducting polymer nanomaterials. Nanomaterials (Basel) 2013; 3(3): 524-49.
[http://dx.doi.org/10.3390/nano3030524] [PMID: 28348348]

[28]   Liu H, Li Q, Zhang S, *et al.* Electrically conductive polymer composites for smart flexible strain sensors: a critical review. J Mater Chem C Mater Opt Electron Devices 2018; 6(45): 12121-41.
[http://dx.doi.org/10.1039/C8TC04079F]

[29]   Wang Y, Liu A, Han Y, Li T. Sensors based on conductive polymers and their composites: a review. Polym Int 2020; 69(1): 7-17.
[http://dx.doi.org/10.1002/pi.5907]

[30]   Lin S, Zhao X, Jiang X, *et al.* Highly stretchable, adaptable, and durable strain sensing based on a bioinspired dynamically cross-linked graphene/polymer composite. Small 2019; 15(19): 1900848.
[http://dx.doi.org/10.1002/smll.201900848] [PMID: 30957404]

[31]   Ehsani M, Rahimi P, Joseph Y. Structure–function relationships of nanocarbon/polymer composites for chemiresistive sensing: a review. Sensors (Basel) 2021; 21(9): 3291.
[http://dx.doi.org/10.3390/s21093291] [PMID: 34068640]

[32]   Chiang CJ, Tsai K-T, Lee Y-H, *et al. In situ* fabrication of conducting polymer composite film as a chemical resistive CO_2 gas sensor. Microelectron Eng 2013; 111: 409-15.
[http://dx.doi.org/10.1016/j.mee.2013.04.014]

[33]   Iijima S. Helical microtubules of graphitic carbon. Nature 1991; 354(6348): 56-8.
[http://dx.doi.org/10.1038/354056a0]

[34]   Iijima S, Ichihashi T. Single-shell carbon nanotubes of 1-nm diameter. Nature 1993; 363(6430): 603-5.
[http://dx.doi.org/10.1038/363603a0]

[35]   Ando Y, Zhao X. Synthesis of carbon nanotubes by arc-discharge method. New Diam Front Carbon Technol 2006; 16.

[36]   Radhakrishnan G, Adams PM, Bernstein LS. Room-temperature deposition of carbon nanomaterials by excimer laser ablation. Thin Solid Films 2006; 515(3): 1142-6.
[http://dx.doi.org/10.1016/j.tsf.2006.07.120]

[37]   Radhakrishnan G, Adams PM, Bernstein LS. Plasma characterization and room temperature growth of carbon nanotubes and nano-onions by excimer laser ablation. Appl Surf Sci 2007; 253(19): 7651-5.
[http://dx.doi.org/10.1016/j.apsusc.2007.02.033]

[38]   Rajura, Efficient chemical vapour deposition and arc discharge system for production of carbon nano-tubes on a gram scale: Review of Scientific Instruments: 90, No 12.
https://aip.scitation.org/doi/abs/10.1063/1.5113850 (accessed Oct. 04, 2021).

[39]   Gajewski S. Purification of single walled carbon nanotubes by thermal gas phase oxidation Diamond and Related Materials 2003; 12(3-7): 816-20.
[http://dx.doi.org/10.1016/S0925-9635(02)00362-X]

[40]   Hou PX, Liu C, Cheng HM. Purification of carbon nanotubes. Carbon 2008; 46(15): 2003-25.
[http://dx.doi.org/10.1016/j.carbon.2008.09.009]

[41]   Rinzler AG, Liu J, Dai H, *et al.* Large-scale purification of single-wall carbon nanotubes: process, product, and characterization. Appl Phys, A Mater Sci Process 1998; 67(1): 29-37.
[http://dx.doi.org/10.1007/s003390050734]

[42]   Xu C, E Flahaut, J Sloan *et.al.* Purification of single-walled carbon nanotubes grown by a chemical vapour deposition (CVD) method. Chem Res Chin Univ 2002; 18: 130-2.

[43]   Dillon AC, Yudasaka M, Dresselhaus MS. Employing Raman spectroscopy to qualitatively evaluate the purity of carbon single-wall nanotube materials. J Nanosci Nanotechnol 2004; 4(7): 691-703.
[http://dx.doi.org/10.1166/jnn.2004.116] [PMID: 15570946]

[44]   Christ J, Aliheidari N, Pötschke P, Ameli A. Bidirectional and stretchable piezoresistive sensors enabled by multimaterial 3d printing of carbon nanotube/thermoplastic polyurethane nanocomposites. Polymers (Basel) 2018; 11(1): 11.
[http://dx.doi.org/10.3390/polym11010011] [PMID: 30959995]

[45]   Robertson DH, Brenner DW, Mintmire JW. Energetics of nanoscale graphitic tubules. Phys Rev B Condens Matter 1992; 45(21): 12592-5.
[http://dx.doi.org/10.1103/PhysRevB.45.12592] [PMID: 10001304]

[46]   Lucas AA, Lambin PH, Smalley RE. On the energetics of tubular fullerenes. J Phys Chem Solids 1993; 54(5): 587-93.
[http://dx.doi.org/10.1016/0022-3697(93)90237-L]

[47]   Bacon R. Growth, Structure, and Properties of Graphite Whiskers. J Appl Phys 1960; 31(2): 283-90.
[http://dx.doi.org/10.1063/1.1735559]

[48]   Qi P, Vermesh O, Grecu M, *et al.* Toward large arrays of multiplex functionalized carbon nanotube sensors for highly sensitive and selective molecular detection. Nano Lett 2003; 3(3): 347-51.
[http://dx.doi.org/10.1021/nl034010k] [PMID: 36517998]

[49]   Sumanasekera GU, Adu CKW, Fang S, Eklund PC. Effects of gas adsorption and collisions on electrical transport in single-walled carbon nanotubes. Phys Rev Lett 2000; 85(5): 1096-9.
[http://dx.doi.org/10.1103/PhysRevLett.85.1096] [PMID: 10991483]

[50]   Kong J, Franklin NR, Zhou C, *et al.* Nanotube molecular wires as chemical sensors. Science 2000; 287(5453): 622-5.
[http://dx.doi.org/10.1126/science.287.5453.622] [PMID: 10649989]

[51]   Zhou C, Kong J, Dai H. Electrical measurements of individual semiconducting single-walled carbon nanotubes of various diameters. Appl Phys Lett 2000; 76(12): 1597-9.
[http://dx.doi.org/10.1063/1.126107]

[52]   Someya T, Small J, Kim P, Nuckolls C, Yardley JT. Alcohol vapor sensors based on single-walled carbon nanotube field effect transistors. Nano Lett 2003; 3(7): 877-81.
[http://dx.doi.org/10.1021/nl034061h]

[53]   Novak JP, Snow ES, Houser EJ, Park D, Stepnowski JL, McGill RA. Nerve agent detection using networks of single-walled carbon nanotubes. Appl Phys Lett 2003; 83(19): 4026-8.
[http://dx.doi.org/10.1063/1.1626265]

[54]   Chang YW, Oh JS, Yoo SH, Choi HH, Yoo KH. Electrically refreshable carbon-nanotube-based gas sensors. Nanotechnology 2007; 18(43): 435504.
[http://dx.doi.org/10.1088/0957-4484/18/43/435504]

[55]   Valentini L, Cantalini C, Armentano I, Kenny JM, Lozzi L, Santucci S. Investigation of the NO[sub 2]

sensitivity properties of multiwalled carbon nanotubes prepared by plasma enhanced chemical vapor deposition. J Vac Sci Technol B 2003; 21(5): 1996-2000.
[http://dx.doi.org/10.1116/1.1599858]

[56]   Li J, Lu Y, Ye Q, Cinke M, Han J, Meyyappan M. Carbon nanotube sensors for gas and organic vapor detection. Nano Lett 2003; 3(7): 929-33.
[http://dx.doi.org/10.1021/nl034220x]

[57]   Suehiro J, Zhou G, Hara M. Fabrication of a carbon nanotube-based gas sensor using dielectrophoresis and its application for ammonia detection by impedance spectroscopy. J Phys D Appl Phys 2003; 36(21): L109-14.
[http://dx.doi.org/10.1088/0022-3727/36/21/L01]

[58]   Li J, Lu Y, Ye Q, Delzeit L, Meyyappan M. A gas sensor array using carbon nanotubes and microfabrication technology. Electrochem Solid-State Lett 2005; 8(11): H100.
[http://dx.doi.org/10.1149/1.2063289]

[59]   Goldoni A, Larciprete R, Petaccia L, Lizzit S. Single-wall carbon nanotube interaction with gases: sample contaminants and environmental monitoring. J Am Chem Soc 2003; 125(37): 11329-33.
[http://dx.doi.org/10.1021/ja034898e] [PMID: 16220955]

[60]   Sippel-Oakley J, Wang HT, Kang BS, *et al.* Carbon nanotube films for room temperature hydrogen sensing. Nanotechnology 2005; 16(10): 2218-21.
[http://dx.doi.org/10.1088/0957-4484/16/10/040] [PMID: 20817998]

[61]   Matranga C, Bockrath B. Hydrogen-bonded and physisorbed CO in single-walled carbon nanotube bundles. J Phys Chem B 2005; 109(11): 4853-64.
[http://dx.doi.org/10.1021/jp0464122] [PMID: 16863139]

[62]   Bindra P, Hazra A. Capacitive gas and vapor sensors using nanomaterials. J Mater Sci Mater Electron 2018; 29(8): 6129-48.
[http://dx.doi.org/10.1007/s10854-018-8606-2]

[63]   Umesh S, Balachandra TC. Carbon nanotube based Surface Acoustic Wave gas sensor for condition monitoring of gas insulated switchgear systems 2015 International Conference on Emerging Research in Electronics, Computer Science and Technology (ICERECT) 2015; 413-7.
[http://dx.doi.org/10.1109/ERECT.2015.7499051]

[64]   Singh SK, Azad P, Akhtar MJ, Kar KK. Improved methanol detection using carbon nanotube-coated carbon fibers integrated with a split-ring resonator-based microwave sensor. ACS Appl Nano Mater 2018; 1(9): 4746-55.
[http://dx.doi.org/10.1021/acsanm.8b00965]

[65]   Zhang T, Mubeen S, Myung NV, Deshusses MA. Recent progress in carbon nanotube-based gas sensors. Nanotechnology 2008; 19(33): 332001.
[http://dx.doi.org/10.1088/0957-4484/19/33/332001] [PMID: 21730614]

[66]   Schroeder V, Savagatrup S, He M, Lin S, Swager TM. Carbon nanotube chemical sensors. Chem Rev 2019; 119(1): 599-663.
[http://dx.doi.org/10.1021/acs.chemrev.8b00340] [PMID: 30226055]

[67]   Ghodrati M, Mir A, Farmani A. Carbon nanotube field effect transistors–based gas sensors.Nanosensors for Smart Cities. Elsevier 2020; pp. 171-83.
[http://dx.doi.org/10.1016/B978-0-12-819870-4.00036-0]

[68]   Kauffman DR, Star A. Carbon nanotube gas and vapor sensors. Angew Chem Int Ed 2008; 47(35): 6550-70.
[http://dx.doi.org/10.1002/anie.200704488] [PMID: 18642264]

[69]   Feng X, Irle S, Witek H, Morokuma K, Vidic R, Borguet E. Sensitivity of ammonia interaction with single-walled carbon nanotube bundles to the presence of defect sites and functionalities. J Am Chem Soc 2005; 127(30): 10533-8.

[http://dx.doi.org/10.1021/ja042998u] [PMID: 16045340]

[70]   Wang R, Zhang D, Sun W, Han Z, Liu C. A novel aluminum-doped carbon nanotubes sensor for carbon monoxide. J Mol Struct THEOCHEM 2007; 806(1-3): 93-7.
[http://dx.doi.org/10.1016/j.theochem.2006.11.012]

[71]   Rushi AD, Gaikwad S, Deshmukh M, Patil H, Bodkhe G, Shirsat MD. Functionalized carbon nanotubes: Facile development of gas sensor platform. AIP Conf Proc 2016; 1728(1): 020164.
[http://dx.doi.org/10.1063/1.4946215]

[72]   Aroutiounian VM. Metal oxide gas sensors decorated with carbon nanotubes. Lith J Phys 2016; 55(4): 4.
[http://dx.doi.org/10.3952/physics.v55i4.3230]

[73]   Morsy M, Yahia IS, Zahran HY, Meng F, Ibrahim M. Portable and battery operated ammonia gas sensor based on cnts/rgo/zno nanocomposite. J Electron Mater 2019; 48(11): 7328-35.
[http://dx.doi.org/10.1007/s11664-019-07550-7]

[74]   Sharma AK, Debnath AK, Aswal DK, Mahajan A. Room temperature ppb level detection of chlorine using peripherally alkoxy substituted phthalocyanine/SWCNTs based chemiresistive sensors. Sens Actuators B Chem 2022; 350: 130870.
[http://dx.doi.org/10.1016/j.snb.2021.130870]

[75]   Suhail MH, Abdullah OG, Kadhim GA. Hydrogen sulfide sensors based on PANI/f-SWCNT polymer nanocomposite thin films prepared by electrochemical polymerization. J Sci Adv Mater Devices 2019; 4(1): 143-9.
[http://dx.doi.org/10.1016/j.jsamd.2018.11.006]

[76]   Liu B, Liu X, Yuan Z, *et al.* A flexible NO_2 gas sensor based on polypyrrole/nitrogen-doped multiwall carbon nanotube operating at room temperature. Sens Actuators B Chem 2019; 295: 86-92.
[http://dx.doi.org/10.1016/j.snb.2019.05.065]

[77]   Farea MA, Mohammed HY, Shirsat SM, *et al.* Hazardous gases sensors based on conducting polymer composites: Review. Chem Phys Lett 2021; 776: 138703.
[http://dx.doi.org/10.1016/j.cplett.2021.138703]

[78]   Star A, Han TR, Joshi V, Gabriel JCP, Grüner G. Nanoelectronic carbon dioxide sensors. Adv Mater 2004; 16(22): 2049-52.
[http://dx.doi.org/10.1002/adma.200400322]

[79]   Sattler KD. 21st Century nanoscience – a handbook: low-dimensional materials and morphologies (volume four).. CRC Press 2020.

[80]   Datta K, Ghosh P, More MA, Shirsat MD, Mulchandani A. Controlled functionalization of single-walled carbon nanotubes for enhanced ammonia sensing: a comparative study. J Phys D Appl Phys 2012; 45(35): 355305.
[http://dx.doi.org/10.1088/0022-3727/45/35/355305]

[81]   Kum M, Joshi K, Chen W, Myung N, Mulchandani A. Biomolecules-carbon nanotubes doped conducting polymer nanocomposites and their sensor application. Talanta 2007; 74(3): 370-5.
[http://dx.doi.org/10.1016/j.talanta.2007.08.047] [PMID: 18371651]

[82]   Chekin F, Gorton L, Tapsobea I. Direct and mediated electrochemistry of peroxidase and its electrocatalysis on a variety of screen-printed carbon electrodes: amperometric hydrogen peroxide and phenols biosensor. Anal Bioanal Chem 2015; 407(2): 439-46.
[http://dx.doi.org/10.1007/s00216-014-8282-x] [PMID: 25374125]

[83]   Ghosh P, Datta K, Mulchandani A, Sonkawade RG, Asokan K, Shirsat MD. A chemiresistive sensor based on conducting polymer/SWNT composite nanofibrillar matrix—effect of 100 MeV O [16] ion irradiation on gas sensing properties. Smart Mater Struct 2013; 22(3): 035004.
[http://dx.doi.org/10.1088/0964-1726/22/3/035004]

[84]   Gao Y, Li L, Tan P, Liu L, Zhang Z. Application of Raman spectroscopy in carbon nanotube-based

polymer composites. Chin Sci Bull 2010; 55(35): 3978-88.
[http://dx.doi.org/10.1007/s11434-010-4100-9]

[85] Wescott JT, Kung P, Maiti A. Conductivity of carbon nanotube polymer composites. Appl Phys Lett 2007; 90(3): 033116.
[http://dx.doi.org/10.1063/1.2432237]

[86] Chen J, Yan L, Song W, Xu D. Interfacial characteristics of carbon nanotube-polymer composites: A review. Compos, Part A Appl Sci Manuf 2018; 114: 149-69.
[http://dx.doi.org/10.1016/j.compositesa.2018.08.021]

[87] Yoon B, Choi SJ, Swager TM, Walsh GF. Switchable single-walled carbon nanotube–polymer composites for co $_2$ sensing. ACS Appl Mater Interfaces 2018; 10(39): 33373-9.
[http://dx.doi.org/10.1021/acsami.8b11689] [PMID: 30229659]

[88] Luo YL, Wei XP, Cao D, Bai RX, Xu F, Chen YS. Polystyrene-block-poly(tert-butyl methacrylate)/multiwall carbon nanotube ternary conducting polymer nanocomposites based on compatibilizers: Preparation, characterization and vapor sensing applications. Mater Des 2015; 87: 149-56.
[http://dx.doi.org/10.1016/j.matdes.2015.08.030]

[89] Choi HH, Lee J, Dong KY, Ju BK, Lee W. Gas Sensing performance of composite materials using conducting polymer/single-walled carbon nanotubes. Macromol Res 2012; 20(2): 143-6.
[http://dx.doi.org/10.1007/s13233-012-0030-5]

[90] Xu G, Li B, Wang X, Luo X. Electrochemical sensor for nitrobenzene based on carbon paste electrode modified with a poly(3,4-ethylenedioxythiophene) and carbon nanotube nanocomposite. Mikrochim Acta 2014; 181(3-4): 463-9.
[http://dx.doi.org/10.1007/s00604-013-1136-y]

[91] Srivastava S, Sharma SS, Kumar S, Agrawal S, Singh M, Vijay YK. Characterization of gas sensing behavior of multi walled carbon nanotube polyaniline composite films. Int J Hydrogen Energy 2009; 34(19): 8444-50.
[http://dx.doi.org/10.1016/j.ijhydene.2009.08.017]

[92] Rajalakshmi K, John SA. Highly sensitive determination of nitrite using FMWCNTs-conducting polymer composite modified electrode. Sens Actuators B Chem 2015; 215: 119-24.
[http://dx.doi.org/10.1016/j.snb.2015.03.050]

[93] Kim T-J, Kim S-D, Min N-K, Pak JJ, Lee C-J, Kim S-W. 'NH $_3$ sensitive chemiresistor sensors using plasma functionalized multiwall carbon nanotubes/conducting polymer composites'. IEEE SENSORS 2008; 208-11.
[http://dx.doi.org/10.1109/ICSENS.2008.4716419]

[94] Lobov IA, Davletkildeev NA, Nesov SN, Sokolov DV, Korusenko PM. Effect of nitrogen atoms in the cnt structure on the gas sensing properties of pani/cnt composite. Appl Sci (Basel) 2022; 12(14): 7169.
[http://dx.doi.org/10.3390/app12147169]

[95] Aliev AE. Bolometric detector on the basis of single-wall carbon nanotube/polymer composite. Infrared Phys Technol 2008; 51(6): 541-5.
[http://dx.doi.org/10.1016/j.infrared.2008.06.003]

[96] Li J, Lee EC. Carbon nanotube/polymer composite electrodes for flexible, attachable electrochemical DNA sensors. Biosens Bioelectron 2015; 71: 414-9.
[http://dx.doi.org/10.1016/j.bios.2015.04.045] [PMID: 25950937]

[97] Salavagione HJ, Díez-Pascual AM, Lázaro E, Vera S, Gómez-Fatou MA. Chemical sensors based on polymer composites with carbon nanotubes and graphene: the role of the polymer. J Mater Chem A Mater Energy Sustain 2014; 2(35): 14289-328.
[http://dx.doi.org/10.1039/C4TA02159B]

[98] Pissis P, Georgousis G, Pandis C, *et al.* Strain and damage sensing in polymer composites and

nanocomposites with conducting fillers. Procedia Eng 2015; 114: 590-7.
[http://dx.doi.org/10.1016/j.proeng.2015.08.109]

[99]   Kanoun O, Bouhamed A, Ramalingame R, Bautista-Quijano JR, Rajendran D, Al-Hamry A. Review on conductive polymer/cnts nanocomposites based flexible and stretchable strain and pressure sensors. Sensors (Basel) 2021; 21(2): 341.
[http://dx.doi.org/10.3390/s21020341] [PMID: 33419047]

CHAPTER 7

Polymer Composites for Automotive Applications

Naveen Veeramani[1], **Prosenjit Ghosh**[1], **Tushar Kanti Das**[2] and **Narayan Chandra Das**[2,*]

[1] *Center for Carbon Fiber and Prepregs, CSIR-National Aerospace Laboratories, Bangalore-560017, India*

[2] *Rubber Technology Centre, Indian Institute of Technology Kharagpur, Kharagpur-721302, India*

Abstract: The last couple of decades have witnessed exceptional advancements in automotives; and the use of polymer composites (PCs) in making different automotive parts has emerged as an integral part of the advancement. Fiber-reinforced PCs offer weight benefits to automotives, thus enhancing fuel economy. Moreover, these composites can be engineered for versatile applications, *e.g.*, interior and exterior body parts. Ease of manufacturing is another advantage of PCs, although several major technical considerations still need to address before engineering these composites for wide-scale acceptance in various automotive applications, especially for exterior body parts. However, PCs are a new class of materials, and developing state-of-the-art manufacturing technology may enhance the comfort and security of modern vehicles. This chapter outlines the utility and recent advances in PCs for various automotive applications. In addition, quality assurance and the advantages of PCs are also given. The potential of PCs for future perspectives is also discussed.

Keywords: Automotive, Fuel economy, Polymer composites.

INTRODUCTION

Recent trends of fuel-saving and minimizing emissions during fabrication and transportation are fuelling interest in manufacturing low-cost, lightweight yet high-performance materials as a substitute for metals. Polymer composites (PCs) are alternative materials with excellent mechanical, processing, and thermal properties that have attracted much attention in replacing metallic components in automotive applications. Current automotive industries are welcoming these replacements [1]. PCs have shared almost 90% global market of composite-based manufacturing sectors. PCs exhibit variable properties depending on the reinforcement *i.e.* organic, inorganic, metallic and polymer matrix. Besides, the property of PCs can be further enhanced by selecting a proper composition, fabri-

* **Corresponding author Narayan Chandra Das:** Rubber Technology Centre, Indian Institute of Technology Kharagpur, Kharagpur-721302, India; E-mail: ncdas@rtc.iitkgp.ac.in

Subhendu Bhandari, Prashant Gupta and Ayan Dey (Eds.)

-cation methods, the density of the reinforcing materials, and its orientation in the polymer matrix [2]. So, PCs are gradually replacing the conventional materials used for the fabrication of various automobile parts. Potential research on PCs for automotive applications has been still ongoing to achieve the following benefits;

• Reduction in the weight of composites which enhance fuel economy and its performance.
• Better quality of the composites and reproducibility in fabrication techniques.
• Improve comforts during ride, *i.e.* reduction in noise and vibration.
• Lower investment cost.
• Acceptable style of vehicles depending on the demand of market.

Other areas are also there that require extensive research to improve the properties for various automotive applications [3].

The mechanical properties of PCs are not only the function of reinforcements but also of polymer matrices. This is because polymers are ductile and during breaking of PCs, reinforcements first get damaged on the application of stress, and then elastic deformation of the polymer supplies shear force to resist the applied stress. Such a transfer of load helps the PCs to combat the applied stress [4]. Besides, the orientation of reinforcement in the polymer matrix is a very important parameter and depending upon the final property requirements, the reinforcing materials are oriented in the polymer matrix. If the reinforcing materials are inclined to the applied stress direction, PCs exhibit the highest strength and modulus, while they show the lowest when directed at the transverse direction. Random distribution of reinforcing agents in the polymer matrix facilitates isotropic behavior of PCs, *i.e.* equal performance in all directions to the applied stress [5]. It is to mention here that most of the PCs used in automotive applications comprise randomly oriented reinforcing materials. These composites are fabricated either by compression molding or injection molding techniques. However, although in the longitudinal direction, reinforcing materials provide high mechanical strength, it is very difficult to construct the composites with reinforcing agents either in the longitudinal or transverse direction [6].

In automotive applications, both thermoplastic and thermosetting polymers are used to fabricate the respective PCs. Thermoplastic PCs are used mainly in making various interior and body parts of automobiles. Automotive industries use various thermoplastic polymers *e.g.*, polypropylene (PP), polycarbonate (PC), poly methyl methacrylate (PMMA), high density polyethylene (HDPE), acrylonitrile butadiene styrene (ABS) copolymer, plasticized polyvinyl chloride (p-PVC), polyethylene terephthalate (PET), polyamide-6 (PA-6), polyamide (PA-6,6) polybutylene terephthalate (PBT), *etc.* to make different components. These polymer resins are common in automotive parts making due to their low cost compared to other thermoplastics [1, 7]. On the other hand, general purpose thermoset polymers used in automotive industries are epoxy resin, polyester, vinyl ester, polyimide, polyurethane resin, phenolic resin, and amino resin. These composites are fabricated through sheet molding techniques where reinforcing materials are randomly oriented in the thermoset matrix. Another important method of fabrication of thermosetting composites is structural reaction injection molding, and this method is specifically predominant for polyimide and polyurethane resins [8, 9].

For manufacturing different parts of automotive industries, the primary reinforcing materials impregnated in the polymer matrix are carbon fiber, glass fiber, Kevlar fiber, natural fiber, various metals, and their alloys. Though other reinforcing agents are used for the fabrication of PCs, these are not widely used for automotive industries [10]. The main benefits of carbon fiber as reinforcement are very low density, high stiffness, and high strength to weight ratio. Though its cost is higher compared to other fibers, these can be compromised during the design and fabrication of PCs [11]. Glass fiber is another suitable reinforcing material used with the polymer matrix. There are various types of glass fibers available in the commercial market, such as E-glass, C-glass, R-glass, S-glass, and T-glass. Among these E and C glass fiber-based polymer composites are mostly used in automotive industries. E-glass fibers provide high electrical resistance, whereas C-glass fiber has good resistance to chemical attacks [12]. Depending upon the applications, different glass fibers are impregnated in polymers to fabricate composites. Though the cost of glass fiber is less than carbon fiber, it has a high density and low stiffness. For this reason, the glass fiber reinforced PCs are heavier and thicker than carbon fiber-based composites [13]. Besides, other reinforcing agents are also impregnated into the polymer matrix to manufacture different parts of automotive bodies, but these have less importance than either carbon or glass fibers. Because of these, research and development sections are gradually moving from metal or metal alloy-based PCs to other reinforcing material-based composites [14]. Fig (**1**) shows the usage of polymer composites in automotives to make various spare parts [15].

Fig. (1). Various applications of polymer composites in automotive [15].

USAGE OF POLYMER COMPOSITES IN AUTOMOTIVE PARTS AND COMPONENTS

The usage of PCs in the automotive industry has increased with time. Metallic components of automotive parts are being replaced by PCs due to their high strength-to-weight ratio. In automotive industries, fabrications of various parts by PCs are one of the safer and economical techniques. In 1953, the first glass fiber-reinforced PC was used in the body of the Chevrolet Corvette [16]. Among others, fiber-reinforced (such as carbon fiber, and glass fiber) PCs are widely used in the fabrication of different parts in automotive industries such as wheels, seats, roofs, mats, energy absorbers, engine covers, dashboards, interior and external panels [17]. The primary aim of using these materials is to reduce the overall weight and cost of the components over metallic parts, hence reducing fuel consumption [18]. The automotive section employs 50% thermoplastic composites and 25% thermoset PCs in the overall composite market. Recently, short fiber, glass fiber and blend of glass and carbon fiber-reinforced thermoplastic composites are widely used in bumper beams, door panels, and cross members. For the above applications, PP is employed as a thermoplastic matrix due to its high toughness, good fatigue, and heat resistance, moderate chemical resistance, and ease of processing to fabricate complex parts [19, 20].

In the recent past, a new technology of PCs has come into the picture, *i.e.*, self-reinforced polymer (SRP) composites in which polymers and fibers used for the fabrication are of the same type. As the polymer and reinforcing materials are of similar type, strong interfacial interaction exists between them. This type of SRP composite is used for various structural applications, such as seat frame, door panel, safety helmet, cover, and shells for luggage [21]. The thermoset PCs are primarily fabricated through a simple compression molding process. The automotive parts fabricated by this process are radiator supports, deck lids, hoods, engine valve covers, oil pans, door pans fender, *etc.* Randomly orientated glass fibers are impregnated in the thermoset polymers such as vinyl ester resin, polyester, and polyimide to fabricate the above automotive composites parts [22]. The advantage of the sheet molding technique is not only to reduce the cost and weight of fabricated parts but also to consolidate the parts easily. For example, the radiator support is replaced by the sheet molding PCs instead of carbon steel [23]. Glass fiber-reinforced epoxy resin composites are used to manufacture leaf springs, which give five times more fatigue resistance than a steel body. The leaf springs fabricated by PCs provide an immediate response to any shock and smooth-riding ability. Besides, PCs leaf springs have more resistance to failure and corrosion resistance [24].

Nowadays, the engine valve guide is prepared by tungsten filler embedded epoxy resin composites instead of a metal valve guide. The replacement of metal valves by PCs not only lowers the weight but also reduces vibration, noise, and the rise in temperature due to friction [25]. High-wear resistance clutch plates are manufactured by hybrid fabric reinforced polyester resin. These PCs clutch plates provide high longevity, less energy consumption during power transmission, and excellent machinability [26]. The hoods of modern cars are typically made up of chopped glass fiber-reinforced thermosetting PCs using the sheet molding process [27]. Several well-known automobile companies are already using carbon-fiber embedded epoxy resin composites in some selected portions. For example, BMW group is fabricating the roof panel of cars with these composites prepared by resin transfer molding. The prepared panels are thicker than conventional steel panels but have less weight, thus providing better stability for design [28].

Carbon fiber-reinforced PCs are widely used in sports cars because of their lightweight, where the cost is not a bar for designing. Previously, the aluminium body panels of racing cars were replaced by glass fiber reinforced polyester composites. But now all the crucial parts of racing cars such as chassis and suspension components are made by carbon fiber reinforced epoxy resin composites due to their lightweight, ease of processing, low curing shrinkage, good mechanical properties, and easy bonding with other parts [29]. Survival cells that save life during any crash event are also prepared by carbon fiber-reinforced epoxy resin [30]. The low-cost semi-structural components such as backup bumper beams, load floors, *etc.* are prepared by randomly oriented wood glass-filled or short glass fiber-reinforced PP or PET. However, recently PP and PET are being replaced by other high-performance polymers such as polyether ether ketone (PEEK) and polyphenylene sulphide (PPS). But, the cost of PEEK or PPS is very high, so the fabricated composites are of limited use [31].

Recently, natural fiber-based PCs have gained interest in manufacturing automotive parts. Mercedes-Benz is fabricating jute-filled epoxy composites for door panels, while door trim panels are prepared by flax reinforced PU by Audi [32, 33]. Although it is relatively easy to fabricate composites using natural fiber in polymer matrix for various automotive applications, it is difficult to fabricate exterior parts with natural fibers as the exterior weather conditions such as humidity, and temperature can affect the characteristic properties of the fabricated parts [34]. So, the exterior body panels of automotive parts are fabricated by glass fiber-filled polymers such as PP, PET, and PBT not affected by the external parameters. Additionally, natural fiber-based PCs exhibit a smooth and glossy surface finish that enhances the aesthetics of the cars [35]. The aramid fiber reinforced PCs are used in the rotating automotive parts where the reduction in friction and dimensional stability is more important than mechanical strength. These aramid-filled PCs are used in upholstery fabrics, body armour, helmets, and vehicle armour [36]. The automotive steering column is very important for the safety of vehicles and fatigue failure of these parts under cyclic loading is crucial to determine its lifetime. For these, the parts are manufactured by carbon fiber-filled PCs through pulforming technique [37]. PMMA based PCs are used in automotive industries where optical transparency is an important parameter, *e.g.*, the rear light lens clusters. These are recently prepared by silica powder-filled PMMA using the simple solution casting technique [38]. The specific usage of different fiber reinforced PCs are discussed in the following sections.

Glass Fiber-Reinforced PCs

Glass fibers are commercially available with different chemical compositions. Mostly, glass fibers are made of silica with a SiO_2 content of 50-60% and oxides of boron, sodium, calcium, iron, and aluminium. The major types of glass fibers are E-glass, C-glass, and S-glass and are normally reinforced with thermosetting or thermoplastic polymers. E-glass means an electrical glass fiber with very good electrical insulating properties. C-glass refers to corrosion/chemical resistance properties. S-glass indicates the fiber with high silica content. E-glass is the most commonly produced continuous glass fiber in bulk quantity [39]. Glass fiber composites are the major contributors in making automotive parts due to their major advantages like low cost of production, corrosion resistant, chemical resistant, lightweight. The important automotive parts being made using glass fiber composites are bumpers, hoods, and casings. As the glass fibers have higher tensile strength, they are also used for making V-belts and timing belts through impregnating glass strings with rubber. Based on the total quantity of material used, glass fiber composites dominate over the other fiber composites for making automotive parts. Another advantage of glass fiber is the abrasion resistance and hence used for making brake pads and clutches. It is easy to add color to the glass fiber composites.

Glass-filled sheet molding compound (SMC) is a well-established technology for the performance properties or moldability of automotive components. The application of SMC in the areas of painted and unpainted automotive componentry includes structural elements, interior and exterior panels, bumpers, fenders, and high-temperature underwood parts. Also, SMCs are so versatile reinforced plastics to make automotive components. Multi-piece steel and composite rear box from high-strength SMC were made by Honda motors, and a first-ever pickup truck of Ridgeline made SMC parts was reported by Meridian Automotive Systems [40]. Recent technologies have made SMCs more robust and versatile to make tailor-made components in automotive industries. Many engineering problems in automotive have been solved after starting using SMC parts. American Composites Manufacturers Association (ACMA) has listed out the major advantages of SMC for making automotive parts as:

- Reduced tool cost (40% lesser tool cost than steel stamping).
- Substantial weight reduction ($\approx$ 20-35% lighter than steel parts).
- Easy manufacturing process through part integration in a single assembly.
- Higher damage resistance from dents and dings compared with aluminum parts.
- Excellent corrosion resistance.
- Improved noise, vibration, and harshness properties (NVH).
- Design flexibility for complex geometries.

Researchers report glass fiber reinforced plastics (GFRP) to be very effective for making automotive products not just because of their low cost and lightweight, but also of their excellent mechanical properties. Barbaz *et al.*, have reported GFRP composite specimens fabricated with two different glass fibers and thermoplastic matrices. The composite specimens were tested for low velocity impact response at an impact energy of 15 J and obtained the response in terms of acceleration-time and force-displacement histories. Interestingly, the specific energy absorption of glass fiber-reinforced thermoplastic composites was higher than steel. GFRP thermoplastic composites not only reduce the weight of automotive parts but also increase the safety of vehicle in low-speed crashes [41]. In recent times, researchers have developed three-dimensional (3D) stitched preforms using E-glass fiber composites with improved delamination (Mode-I) properties. Nano-fillers like single-walled, multi-walled carbon nanotube are embedded into the glass fiber structure during the preform stitching stage to improve the thermo-mechanical properties of the E-glass fiber composites [42].

In the recent past, many automotive industries started making their components using glass fiber-based SMC. For example, Meridian Automotive Systems (USA) is one of the major companies to use SMC extensively in their exteriors. They have a huge molding press with a capacity ranging from 400 to 4000 tons, to make heavy truck hoods and fenders in a single piece. Most of their SMC products are used in passenger cars, light trucks, deck lids, roof panels, hood and door assemblies, spoilers, back panels, fenders, wheelhouses, bumpers, grills, and firewalls [43]. Automotive parts made of SMC are also used for high-temperature applications. AV-2016, a high-temperature resin system developed by IDI composites Ltd. is used for valve covers with temperature resistance up to 149°C and improved tensile strength and modulus. A high-temperature SMC valve cover used in heavy truck diesel engines is developed by Meridian Automotive Systems. Some of the other automotive companies which use thermoplastic composites with a continuous reinforcement of glass fiber are Daimler Chrysler sports cars (integrated composite seat), Porsche (transverse support beam), and BMW (bumper structure). Some of the major automotive parts, which are made by familiar automotive companies, using glass fiber composites are listed below [4]:

  i. Safety seat and side door structures - BMW
 ii. Floor panel - Nissan
iii. Transverse support beam, door, dashboard - Porsche
 iv. Bumper - BMW
  v. Triangle joint - Audi
 vi. Integral seat - Daimler Chrysler

Carbon Fiber Composites

Carbon fiber is an amazing material because of its excellent strength-to-weight ratio that makes it roughly five times stronger and four times lighter than steel. Carbon fiber-reinforced composite (CFRP) is an ideal choice for the automotive market as it demands high volume production of the components. The global demand for carbon fiber increased rapidly with time. A market survey has predicted that the global demand for carbon fiber composites is expected to reach a massive 1,17,000 metric tons by the end of the year 2022 [44, 45]. CFRP has many desirable properties, such as high tensile and compressive strength, specific stiffness, fatigue resistance, low thermal expansion coefficient, and easiness for making complex shapes which make them more suitable for making automotive components. Other advantages of CFRP include weight reduction, part integration, crashworthiness, toughness, durability, and aesthetic appealing which are desirable properties for automotive applications [46]. Fig. (**2**) shows the advantages of CFRP composites [47].

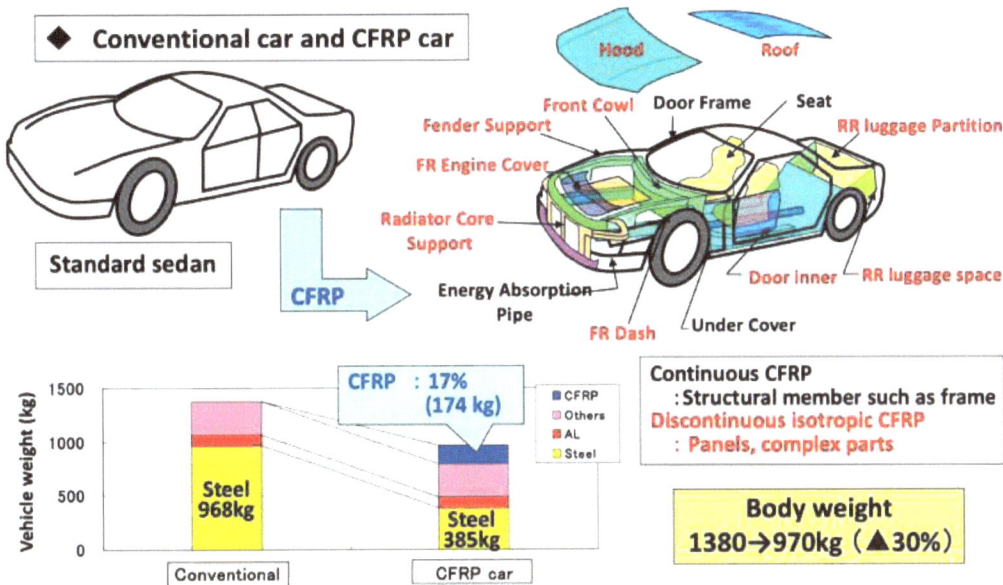

Fig. (2). Weight reduction of automotive using CFRP composites [47].

Composite industries have developed various resin systems and prepregs to make CFRP-based automotive components. Carbon fiber and epoxy-based commercial prepregs system known as "VORAFUSE" is recently developed by Dow Automotive Systems. These prepregs are easy to handle and have improved cycle time during compression molding to make composite structures. The potential application of VORAFUSE in the automotive sector includes, but is not limited to, hood inners, door inners, roof rails, and A, B, C pillars [48]. In recent years, fast curing agents and new preform materials were developed for making CFRP components. Resin transfer molding and out-of-autoclave methods have become attractive options for the mass production of CFRP composite parts for automotive applications.

European Union framework program for automotive applications is working on a project called HIVOCOMP to develop advanced materials enabling high-volume road transport applications of lightweight structural composite parts. In this project, two materials system will be developed with an aim of cost-effective and higher volume production of CFRP components for automotive applications. The two material systems are advanced PU thermoset matrix materials with reduced curing cycle time and improved mechanical performance and thermoplastic PP and PA 6-based self-reinforced polymer composites with continuous carbon fiber reinforcements for reduced cycle times and increased toughness compared to the existing thermoset and thermoplastic systems. The key engineering problems, such as the reduction of resin viscosity, and improved reaction kinetics will be addressed to achieve reduced curing cycle time. The potential automotive applications of HIVOCOMP-based CFRP composite materials are listed below [49]:

   i. Inner bonnet
  ii. Rear seat back panel
 iii. Cross member/crash box
  iv. Floor structure
   v. B-Pillar reinforcement
  vi. Suitcase

Some of the top applications of CFRP in recent times include [50]:

   i. Rear suspension knuckle developed by Ford Motor Co. (Dearborn, Mich., U.S.) using SMC and prepregs.
  ii. Advanced SMC steering knuckle developed by Marelli (Corbetta, Italy).
 iii. Hybrid carbon fiber/aluminum suspension knuckle by Saint Jean Industries (Saint Jean D'Ardières, France) using prepregs from Hexcel (Les Avenières, France).
  iv. Carbon fiber/epoxy suspension links press-formed over aluminum by Shape Machining Ltd. (Oxford shire, U.K.).
   v. CFRP stabilizer bars by IFA Composite (Haldensleben, Germany).
  vi. CFRP wishbones molded in 90 seconds using recycled carbon fiber and the RACETRAK process, developed by Williams Advanced Engineering

(Oxfordshire, U.K.).

vii. Arch-shaped, multifunctional unidirectional glass fiber/epoxy front axle "blade" that incorporates suspension, anti-vibration/noise and anti-roll

viii. CFRP output shaft developed by Dynexa (Laudenbach, Germany)

ix. CFRP-ultra-lightweight SMC (<1.0 g/cc): Polynt Composites (Scanzorosciate, Italy, Aliancys (Schaffhausen, Switzerland) and CSP VICTALL (Tanshan, China)

x. Polynt-RE Carbon recycled fiber SMC: Polynt Composites (Scanzorosciate, Italy

xi. locally-reinforced and co-molded chopped carbon fiber SMC with patches of SMC made with carbon fiber 0-degree/90-degree non-crimp fabric (NCF): Magna International (Aurora, Ontario, Canada) and Ford Motor Company

Some of the disadvantage of CFRP composites are:

i. Cost of production
ii. Complex process technologies
iii. Recyclability

Hybrid Composites

Hybrid composites are produced from matrices reinforced with two or more fibers. Synergic effects of different fibers are expected to improve the overall properties of the final composite product. For example, hemp/glass fiber hybrid PP composites reinforced with hemp (25%) /glass fibers (15%) have resulted in flexural strength and flexural modulus of 101 MPa and 5.5 GPa respectively. The impact strength and water absorption properties of these hybrid composites are also improved [51]. Epoxy/carbon fiber/flax fiber-based hybrid composite has decreased the average weight of the composite material by 17.98%. Also, improved interlinear shear strength of 4.9 MPa and hardness of 77.66 HRC were reported [52]. A hybrid composite made of epoxy/banana fiber (27%)/jute fiber (9%) has shown an improved tensile strength of 29.47 MPa. Similarly, epoxy/coconut sheath (21.5%)/jute (15.5%) based composite has given a compressive strength of 33.87 MPa [53]. The most common hybrid composite combinations are carbon/aramid-reinforced epoxy for combining tensile strength and impact resistance, and glass/carbon fiber reinforced epoxy to make high strength and cost-effective composite. Hybrid composites are used by automotive industries for making many exterior and interior parts due to their easy manufacturing process and a wide range of fabrication processes. Researchers develop specific hybrid composites wherever there is a requirement of multi-functional activity. Various types of hybrid composites are being developed for automotive applications, such as eco-friendly automotive anti-roll bar, automotive piston application, low-velocity impact brake friction materials [2].

Literature also reports on nanocomposites developed for automotive applications. PP/PS/graphene nano-platelets based nanocomposites were developed by embedding with polystyrene-*block*-poly(ethylene-*ran*-butylene)-*block*-polystyrene (SEBS). The graphene nano-platelets were added reportedly to increase the tensile modulus, enthalpy of crystallization, and decrease toughness. The addition of SEBS has improved the toughness and tensile elongation, which are desirable properties for automotive applications [54]. Some of the major applications of lightweight hybrid composites by automotive industries use are listed below [55]:

  i. Air intake manifold - TOYOTA
 ii. Charge-air lines – MERCEDES BENZ
iii. Oil pan - FORD
 iv. Cross beam - BMW
  v. Lower bumper stiffener – GM Motors
 vi. Wheel rim - SMART
vii. Online body panel - MERCEDES BENZ
viii. Structural inserts - PEUGEOT
 ix. Seat - HYUNDAI

Natural Fiber Composites

Natural fibers are available in abundance in nature and are easy to get. Composites made of these fibers have some excellent properties like high strength, specific stiffness, and low cost per unit volume. Natural fiber composites (NFC) also have some unique features over synthetic fibers, such as biodegradability, non-toxic, pollution-free, recyclability [56]. NFCs are broadly classified into wood-based and non-wood-based composites. Further, the reinforcement length is varied as short fiber and long fiber based on the final applications. Bio-composites, reinforced with natural fibers, are extensively used in the automotive industry to produce lightweight components with good mechanical properties, reduced CO_2 emissions, and improved fuel efficiency. Overall, these natural composites reduce 30% of the weight and 20% cost of production of automotive parts. NFCs are used for automotive parts by many European countries especially, Germany. The extent of usage of different NFC in German automotive industries is listed below [57]:

  i. Flax = 64.2%
 ii. Jute / Kenaf = 11.2%
iii. Hemp = 9.5%
 iv. Sisal = 7.3%
  v. Other = 7.9%

The global NFC market is predicted to grow by 11.7% from 6.31 billion USD to 14.6 billion USD between 2019 to 2027. As carbon and glass fiber composites have the concerns like recyclability and biodegradability, NFCs gained popularity in recent times. The major advantages of NFCs are high durability, cost-effectiveness, low density, lightweight, and low cost per unit volume. Some of the natural fibers used to make composite are jute, coir, cotton, hemp, bagasse, kenaf, flax, sisal, and bamboo. Another major advantage of natural fibers is that they can be used with both thermoplastic and thermoset resins to make composites.

NFCs are extensively used in automotive industries to make dashboards, seat backs, door panels, headliners, truck liners, railing, decking, windows, and frames. A recent market survey [58] suggests that the total global consumption of NFC for automatic industries is 28.2%, which is very significant. Among the global automotive sectors, European automotive industries are the major consumers of NFC. Typical usage of various NFC by the European automotive industries is listed below [59]:

  i. Wood = 38%
 ii. Cotton = 25%
iii. Flax = 19%
 iv. Kenaf = 8%
  v. Others (Jute, coir, sisal and abaca) = 7%
 vi. Hemp = 5%

Many literature reports are available on automotive composites made with NFCs. Sisal fiber-based composites are often used to make automobile interiors and upholstery in furniture due to their good tribological properties. A tensile strength value of 12.5 MPa was reported for polyester composites reinforced with 6 mm long sisal fibers [60-62]. Hemps fibers are very effective in making PCs. Typical mechanical properties of hemp fibers are given in Table **1**.

Table 1. Typical mechanical properties of hemp fibres.

Sl. No.	Properties	Values
1	Ultimate length (mm)	8.3 – 14
2	Ultimate diameter (µm)	17 – 23

(Table 1) cont.....

Sl. No.	Properties	Values
3	Aspect ratio (length/diameter)	549
4	Specific apparent density (gravity)	1500
5	Micro-fibril angle (θ)	6.2
6	Moisture content (%)	12
7	Cellulose content (%)	90
8	Tensile strength (MPa)	310 – 750
9	Specific tensile strength (MPa)	210 – 510
10	Young's modulus (GPa)	30 – 60
11	Specific Young's modulus (GPa)	20 – 41
12	Failure strain (%)	2 – 4

Hemp fiber composite with PP matrix has reportedly shown a 52% increase of specific strength compared to that of glass fiber reinforcement [63]. Kenaf fiber and polylactic acid matrix-based composites have resulted in 223 MPa tensile strength and 254 MPa flexural strength [64]. Flax fiber/PP composites have increased stiffness, damping ratio, and high sound absorption properties [65]. Rice husk (5wt.%)/PU-based composite has reportedly enhanced the sound absorption characteristics [66]. Palm fibers improve the fiber-matrix interaction, luffa fibers improve the mechanical properties and water absorption characteristics of the composites, cotton fiber improves the energy absorption and load carrying capacities. The suitable manufacturing process of NFCs is listed in Table **2** [67]. However, NFCs have limitations regarding their reinforcements. Natural fibers are sensitive to moisture, resulting in weak bonding with the polymer matrices. The quality of natural fibers, consistency, and processing temperature limits also matters [68]. In recent times, natural fibers are now being demonstrated in body panels and crash structures for motorsports, which has long been the proving ground before wider adoption. Porsche motorsport partnered with Ford Motors racing team and Bcomp (Fribourg, Switzerland) and they are developing a technology called *Cayman* 718 GT4 CS MR featuring a full natural fiber composite bodywork. The car was premiered at the 24 Hours Nürburgring race in September 2020. They have already launched the Porsche *Cayman* 718 GT4 CS in 2019 with natural fiber doors and rear wings. This product is the first to replace carbon fiber with natural fiber in serial production composites for motorsports [50].

QUALITY ASSURANCE OF PCS FOR AUTOMOTIVE APPLICATIONS

PCs provide automotive manufacturers the flexibility to match the conditions during operation. In fact, these composites do not require the performance criteria of designs to traditional standards [69]. However, the structure of automotive PCs is complex as these are fabricated by combining layers of dissimilar materials. The failure of these composites is also complex and often requires special techniques to reduce defects and damage happening either during their fabrication or service. Employing incorrect process parameters during fabrication introduce defects in the composites, while in-service environment, impact, and irregular stresses of variable magnitude due to extreme operating conditions result in damages to the composite structure [70-72]. The failure mechanism of PCs is also different from that of metals and different techniques are required to determine damage to these composites.

Table 2. Manufacturing process of various fiber reinforced PCs.

Fiber	Matrix	Application	Manufacturing Method
Carbon	Epoxy, PP, PEEK	Light weight automotive parts, fuel cells	Injection molding, filament winding, resin transfer molding (RTM)
Graphene	Epoxy, PS, Polyaniline	Wind turbines, Aircraft/ Automotive parts	Chemical vapour deposition, pultrusion, hand/spray lay-up
Sisal	Epoxy, PP, PS	Automobile body parts	Hand lay-up, compression molding
Hemp	PE, PP, PU	Automotive, furniture	RTM, compression molding
Kenaf	Epoxy, PP, PLA	Automotive parts, bearings	Compression molding, pultrusion,
Flax	Epoxy, Polyester, PP	Structural, textile	RTM, hand lay-up, vacuum infusion
Ramie	PP, PLA, Polyolefin	Bulletproof vests, socket prosthesis	Extrusion with injection molding
Rice Husk	PE, PU	Automotive structures, window/ door frames	Compression/injection molding
Jute	PP, Polyester	Door panels, roofing	Hand lay-up, compression/injection molding
Coir	PE, PP, Polyester	Automotive structural components, roofing sheets	Extrusion/injection molding

Damage in PCs generally occurs through fiber/matrix delamination, fiber pull-outs, fiber breakage, fatigue damage, resin-rich regions, voids, and foreign material inclusions. Among these, void is a serious manufacturing defect that causes undesirable loss in mechanical properties. Voids may originate during manufacturing likely because of air entrapment, leakage in vacuum bags or poor vacuum source, low quality or high-viscosity resin, and formation of by-products or gas bubbles during the curing cycle of the laminate [73, 74]. Voids have detrimental effects on thermal and mechanical properties of the composites and are often responsible for inter-laminar shear failure. Voids also reduce the stability and durability of the composites. The allowable limit of void content in a composite structure is largely application specific *e.g.*, the void content should not exceed 1.5% in dynamic aerospace structures like helicopters, although a void content up to 6% is often accepted in applications under a steady and static load [75]. However, any flaw in a composite structure is not intended and is detrimental to the performance of the component, since such a flaw becomes a defect when that adversely affects the performance of the composite structure [76].

Quality analysis of composites, especially automotive composites, is very essential and important than other structural materials like metals or plastics because there is no accepted database for composites for various engineering applications. In addition, automotive composites experience numerous unsteady deformations, including the adverse effects of extreme environment and temperature during service. Hence, there is an obvious danger in selecting and specifying a PC for automotive application solely based on technical datasheets obtained from material suppliers, material databases, or even handbooks because the reported values of properties *e.g.*, mechanical are generated from standard tests conducted in a laboratory under standard test conditions [77]. Majority of the automotive parts experience some mechanical loading during service, thus the mechanical properties of PCs are very useful for quality assurance. Mechanical properties such as tensile strength, tensile modulus, elongation, impact strength provide valuable guidelines while selecting a PC for specific applications in automotive. For this purpose, a thorough understanding of mechanical properties and the procedure to measure those properties is very much required to predict the performance and life of the PCs. The following sections will discuss the testing method and importance of some of the important mechanical testing of PCs, *e.g.*, tensile, flexural, impact, *etc.*

Tensile Properties

Tensile test is a destructive testing carried out on a PC specimen to acquire information about in-service drawbacks or failure during performance under a standard set of operating conditions. ASTM D3039 is a standard test procedure for measuring the tensile force required to break a PC specimen. The elongation or stretching of the specimen to that breaking point is also noted in this test. Tensile strength is calculated from the maximum force that the PC can withstand without fracture, while tensile modulus is determined from the tensile stress-strain curve. The specimen for tensile testing can be prepared in many different ways like compression molding, injection molding, or by machining of the composite sheet, plat, slab, *etc.* The dimension of the test specimen also varies depending on the composite type (refer to ASTM book of standards) *e.g.*, for cured matrix resin or reinforced thermoset plastics, dumbbell or dog bone shape samples are prepared following ASTM D 638. Standard conditioning procedures are used to condition the specimen before the test. It is noteworthy that the tensile properties of some plastics and reinforced composites may change considerably with the change in temperature. For those materials, tensile test should be carried out in the standard laboratory atmosphere of 23 ± 2 °C and $50\pm5\%$ relative humidity (refer to procedure A of ASTM D618). The speed of testing depends on the PC type; 5 mm/min (0.2 in./min) is the most frequently used one, although ASTM D638 specifies four different testing speeds. Tensile values provide useful guidelines to specify the PC material, to design parts, and for quality control check of materials.

Flexural Properties

Flexural properties of PCs are very important for automotive applications. Flexural test measures the force required to bend the composite beam under different bending forces like three-point or four-point, applied perpendicular to its longitudinal axis. Flexural strength value is often used to select materials for automotive parts that will support loads without flexing. Also, flexural modulus gives an indication of material stiffness when flexed. Thus, the stress-strain behavior of PCs in flexure has paramount importance to design engineers of automotive parts. Flexural stress includes both tensile and compressive stresses during flexural loading. Thus, flexural properties are more relevant for design and specification purposes than the tensile properties for parts used as a beam. The maximum stress and strain that occur at the outside surface of the test beam measure the flexural properties of the composite. Preparing flexural specimens is comparatively easy and the residual strain can be eliminated. More importantly, at small strain, sufficiently large deformation happens during flexural test, which helps in accurate measurement. Alignment of specimen is also simple for this test *e.g.*, in three-point bending, the specimen (a bar of rectangular cross-section) lies on a support span and the load is applied to the centre by the loading nose midway between the support producing three-point bending at a specified rate. The other way of aligning the specimen is a four-point bending method in which the test bar rests on two supports, and loadings are done at two points by two loading noses. Here, two load points are equally spaced from their adjacent support points. The three-point bending test is useful for quality control and specification of PCs that break at comparatively small deflections. Four-point loading is used if the composite does not fail at the maximum stress under the three-point loading system. However, both methods are very useful to determine flexural properties of PCs. The parameters for the flexural test are the support span, strain rate (*i.e.* the speed of the loading), and the maximum deflection for the test. Selection of these parameters is done based on the test specimen thickness and is defined in ASTM D790 or ISO 178.

Impact/ Toughness Properties

Toughness is the ability of a material to absorb energy without rupture. The stress-strain curve of most high-performance composites, *e.g.*, automotive composites, are linear, indicating brittle behavior like ceramics which is in sharp contrast to most metals that exhibit plastic deformation before rupture. This implies high-performance composite materials are essentially elastic to failure like ceramics and can't absorb energy to show plastic deformation like metals [6]. However, energy absorption in ceramics and high-performance composites occurs through the spreading of localized impact energy into a high-volume cone of fractured material. These composites absorb energy by a controlled (fracture) disintegration process, that the impact energy is converted into surface energy, resulting in efficient energy dissipation. According to Strong [78], composites exhibit two toughness behaviors, *e.g.*, equilibrium toughness and impact toughness. The equilibrium toughness is significant during the tensile test at a very low strain rate. This toughness is determined from the area under the stress-strain curve. Impact toughness assumes more importance when a sudden impact is applied to a composite body, hence it is more relevant for automotive composites. ASTM D256 includes Izod and Charpy impact test while ASTM D5628 suggests a falling dart test. However, fracture toughness of laminar composites is best understood by Mode-I fracture toughness testing as per ASTM D5528-13 using a double cantilever beam.

Non-Destructive Testing (NDT)

Non-destructive testing has emerged as an efficient method of PC testing without destroying the structure. It has the advantage that the composite structure or component can still be used after testing, *i.e.* NDT does not harm the structure or parts. NDT employs non-invasive techniques such as ultrasonic testing, radiography, thermography, and microwave and eddy current to determine the integrity of the structure and assures the performance of the part at its highest operating capacity within the specified time limits [69]. Although NDT is a popular in-service inspection technique, it is also used for quality checking of raw materials and reliability of the manufacturing processes. The change in energy signal through or onto the part of the composite under inspection is utilized in NDT techniques. Ultrasonic testing uses acoustic waves in the range of 20-100 MHz and is a very popular NDT for PCs, *e.g.*, carbon/epoxy composites because the size of common defects in these composites matches the wavelength of acoustic waves in the frequency range. Ultrasonic C-scan and A-scan are very sensitive to delamination and disbands in composites. C-scan can detect the size, location, and depth of defects. Although the X-ray technique is a useful tool for metal matrix composites, the same does not apply to polymer matrix composites because most of them are nearly transparent to X-ray. Infrared (IR) thermogram is also a useful NDT technique for PCs to identify their internal flaws. Eddy current technique applies to conductive PCs, whereas surface defects or irregularities in composites are inspected by visual inspection only.

Flammability Testing

Flammability testing has gained practical importance in new product development as well as quality assurance of existing products. This testing is aimed at protecting the public and limit liability through prior experiments. The flammability of PCs mainly depends on the resin content, as fibrous reinforcements used in PCs are more resistant to fire. The properties like the ease of ignition, flame spread, emission of smoke, heat release rate, ease of extinguishing, *etc.* are the characterizing features of the burning of composites. According to Strong (2008) [78], three different classes of flammability tests exist:

i. Official tests for official requirements (ISO 5660/ASTM E1354, ASTM E162, ASTM E662).
ii. Laboratory tests for product development/improvement.
iii. Full scale tests for simulating actual condition.

Official tests are useful in developing combustion models for PCs. ASTM E162 and ASTM E662 are the standards followed for flammability testing of PC-made automotive seats, panels, walls, partitions, and ceilings. Laboratory tests are done for product design and development purposes and are not suitable for ranking the combustion characteristics of different composites. Limited oxygen index (LOI) is a commonly employed ignition test (refer to ASTM D2863) along with the vertical burn test (please see ASTM D568/ASTM D3801), and the horizontal burn test (ASTM D635). LOI indicates the minimum oxygen content required for continuous burning of the PC specimen, the higher the LOI lower is the flammability. Full-scale test determines the performance of the composite under actual combustion conditions [77].

ADVANTAGES OF PCS OVER METAL STRUCTURES

Metal matrix composites (MMC) are gradually being replaced by polymer matrix composites (PMC) for making automotive parts due to their several advantages, such as:

i. Improved processing
ii. Density control

iii. High specific strength to weight ratio
iv. High stiffness to weight ratio
v. Improved fatigue strength
vi. Thermal conductivity
vii. Flame retardancy
viii. Abrasion resistance
ix. Fatigue resistance

PMCs or PCs have an excellent combination of properties such as low density, high stiffness and strength, dimensional stability and excellent corrosion resistance for which they are extensively used in structural applications for all modes of transport *via* land, sea and air. The primary advantages of PCs over other materials are their high specific strength and high specific modulus. Recent developments in PCs make them well suited for UV-resistant and high-temperature applications due to their low cost, lightweight, high strength, and easy processing ability compared to the conventional MMCs. Also, the PCs have higher chemicals resistance when compared to their metal counterparts. Unlike metals, PC parts do not require any post-treatment finishing efforts. These advantages are discussed in some details in the following sub-sections.

High Strength and Low Density

PCs can be even ten times lighter than typical metals and are radar-absorbent as well as thermally and electrically insulating material. Fiber-reinforced PCs with exceptionally high specific strengths and moduli utilize the low-density fiber and matrix properties. In order to obtain effective strengthening and stiffening of the composite material, a critical fiber length is to be used during fabrication. The critical fiber length (l_c) is calculated as per the expression given in equation 1. Most of the glass and CFRP composites have a critical length of around 1 mm. Typically, the value of critical length varies between 20 to 150 times of the fiber diameter [79].

$$l_c = \frac{\sigma_f \times d}{2 \times \tau_c}$$

Where,

σ_f = Tensile strength of the fiber

d = Diameter of the fiber

τ_c = Fiber-matrix bond strength / shear yield strength of matrix

Due to their high strength to weight ratio, carbon-carbon composites are used in high-performance automobiles for hot-pressing molds. PCs are useful for making automobile tires that would be lighter and recyclable. PCs have relatively low density compared to metal and ceramic matrix composites. The density of PCs mainly depends on the fiber volume fraction. Typical tensile strength and density values of widely used fibers and matrices are given in Table **3**.

Fuel Efficiency

The major criterion for improving the fuel efficiency of a vehicle is to reduce its mass. PCs are now widely used by automotive companies to utilize their lightweight benefits that ultimately improve fuel efficiency. Development of next generation energy-efficient vehicles mainly depends on the usage of PCs in automotive components. A 10% reduction in vehicle weight can result in a 6%-

Table 3. Typical mechanical properties of widely used fibres and matrices.

Material	Tensile Strength (GPa)	Tensile Modulus (GPa)	Density (g/cc)
Glass fibre (E-glass)	3.4	72.3	2.58
Standard modulus Carbon fibre	3.5	230	1.80
Intermediate modulus Carbon fibre	5.5	285	1.80
High modulus Carbon fibre	4.5	400	1.80
Epoxy resin	0.1	2.40	1.14

8% fuel economy improvement. In recent times, novel PCs have contributed significantly to the automotive industries mainly because the weight of these composites is 50% lesser compared to the components made of other materials. This contributes to a 25-35% improvement in the fuel economy. The U.S. government has set a target of improving corporate average fuel economy (CAFÉ standards) from a value of around 40 (in 2020) mileage per gallon of fuel (mpg) to 54.5 mpg by 2025.

As PCs have a very high strength-to-weight ratio, these occupy ~50% of a car volume but only 10% by weight. This drastic weight reduction results in lesser strain energy on the engine, improved fuel mileage, and also reduced flue gas emissions. Thus, the automotive parts made of PCs result in lighter, safer, and fuel-efficient vehicles. According to Oak Ridge National Laboratory (ORNL, USA), PCs parts in automotive do not rust or corrode like metal parts, hence can significantly increase the fuel economy by reducing the overall weight of the vehicle as high as 60% [80]. PCs are used in vehicles interiors such as dashboards, control switches, instrument panels, and cup holders as well as various exterior applications like the design of body panels, fuel tanks, fuel lines, windows, heal lamps, bumpers, and taillights. PCs are also used in housings for electronics and braking systems. PC parts protect the components from abrasion and heat damage, extend their reliability and life span, thereby increasing the fuel economy and reducing CO_2 emissions.

Processing Techniques

Most of the PCs used for automotive parts are produced by conventional manufacturing processes such as resin transfer molding, vacuum-assisted resin transfer molding, extrusion, compression molding. However, there are few novel processing techniques such as out-of-autoclave (OAA), and rapid curing are introduced recently. Researchers and manufacturers are putting continuous efforts

Table 4. Various manufacturing techniques for polymer matrix composites.

Manufacturing Process	Polymer Composites Manufactured
Resin transfer molding	Sisal fiber – Polyester composites, Phenoxy nanocomposites, Carbon fiber reinforced composites
Injection molding	Glass fiber reinforced composites, Sisal-Glass fiber hybrid composites, Polypropylene single - polymer composites
Extrusion	Polyethylene- short glass fiber composites, Kenaf – high density polyethylene (HDPE) polymer composites,
Pultrusion	Thermoplastic composites, Natural fiber reinforced composites, Glass fiber / UV cured polyester composites
Compression molding	Carbon fiber – epoxy composites, Sugarcane bagasse cellulose / HDPE composites, Carbon/ PEEK randomly oriented composites,
Filament winding	E-glass fiber - epoxy tubes, MWCNT reinforced composites, Kevlar fiber – epoxy composites
Prepreg tape lay-up	Carbon fiber / Triple-A polyimide composites, Glass fiber – polypropylene composites

into finding new processing techniques that combine all the steps like resin transfer, fiber orientation, rapid curing, and post-curing to get an all-in-one manufacturing technique. These novel methods will help in producing PCs in bulk quantities for making automobile components. PCs have the potential for rapid process cycles, dimensional stability, and lower thermal expansion properties, which are desirable for the ease of processing. ~50% of thermoplastic and 24% of thermoset plastics are accountable for the total consumption of PCs in the automobile sector [81]. Glass fiber composites are the first choice materials for processing due to their fast cycle time, ability to part integration, and low cost. However, high-speed manufacturing processes with cost-effectiveness, *in situ* repairing, and functional requirements are needed to develop for new generation PCs. The major processing methods being used for making polymer composites are listed in Table **4** [82].

Cost Effectiveness

PCs are cost effective when they are produced in large quantities. Machined polymer and their composites give huge cost benefits compared to metals. PCs provide several advantages over metals for bearings and wear applications due to their low frictional properties. Carbon fibers are strong and relatively stiff and provide a low-density reinforcement; however, they are expensive. Glass fibers are cheaper, but have low stiffness. Carbon fibers have low density and provide higher stiffness. However, the cost of production is relatively higher for carbon fibers. Bulk production and advanced manufacturing techniques will reduce the production cost significantly. Hybrid composite is another option for reducing the production cost of PCs. Glass or aramid fibers are usually mixed with other expensive fibers and matrix, which brings down the cost.

DEMAND AND SUPPLY OF PCS TO AUTOMOTIVE SECTORS

Small passenger cars comprise more composite parts compared to large automobiles based on the total weight of the body. Still now, most of the non-structural parts of automotive are fabricated by PCs. The automotive parts are mostly prepared by either glass or carbon fibers reinforced polymers and are processed by a simple sheet molding technique. The primary problem with carbon fiber is that most synthesized carbon fiber is consumed by the aerospace industry. As a result, the automotive industry demands advanced carbon fiber technology that can produce low-cost carbon fiber with a high yield. The high strength to weight ratio of PCs demands its applications in every engineering field. Conventional materials are gradually substituted with PCs in many automotive industries. Research and development on new materials and their performance analysis for automotive applications are continuously going on. For this purpose, lighter weight polymers, a blend of polymers, and biodegradable polymers are becoming the key point of research for manufacturing PCs for the automotive industry. The research aims at fabricating composites that are easy to handle, have low-cost and long lifetime, and are environmentally friendly. As the day's progress, new technologies are also being developed for the low cost manufacturing of PCs. These development processes are not only fulfilling the demands of the automotive industry but also supplying PCs with improved properties and advanced design features [83, 84]. The efficient tooling and assembling of parts for the consolidation of composites may reduce the cost. Hence, the fabrication of PCs for automotive industries requires long life durability data, proper design tools, ease of joining methods, and non-destructive testing methods for their inspection. Fig (3) shows the current market share of FRP composites based on their applications [85].

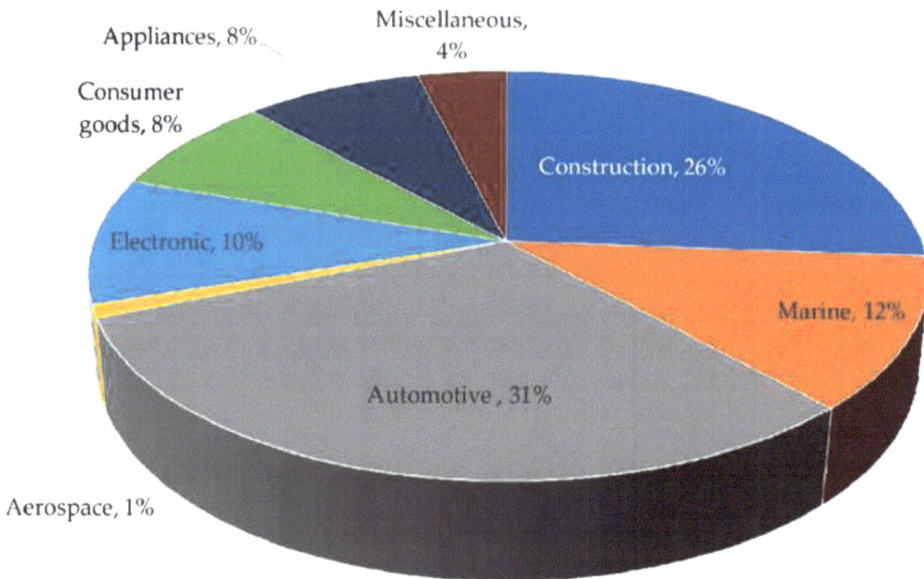

Fig. (3). Market share of FRP composites according to applications in various industrial sectors [85].

FUTURE SCOPE OF PCS IN AUTOMOTIVE

The demand for PCs in the automotive industry has increased with time. A market survey predicts that the global automotive composites market is projected to grow from 5.4 billion $ to 9.3 billion $ between the years 2020 to 2025. This is a very significant growth of 11.5%, despite the reduced demand for automotive composites in recent years (2020-21) due to COVID-19. The survey also estimated the key facts on PCs usage in the automotive sector in the near future [86].

- Glass fibers will be the most consumed reinforcement compared to other fibers for automotive composites in terms of cost. This is because of the superior properties of glass fiber composites, such as stability, flexibility, durability, lightweight, and resistance to heat and moisture. The glass fiber composites are used for making various automotive components such as the underbody, deck lids, front end modules, engine cover, bumper beams, air duct, and others.

- Exterior parts are the major application of automotive composites in terms of cost and volume. The exterior parts include a fender, door panels, hood, and others that impart rigidity and provide minimum risk in case of accidents.
- In terms of cost and volume, non-electric vehicles will be the largest consumers of automotive composites. Major automotive companies like BMW, Audi, Porsche, Volkswagen, and FIAT are using PCs in their high-end non-electric vehicles. CFRP assembly carrier was developed by Porsche for their GT3 cup II model. Fiat Chrysler and BMW use CFRP composites and glass-reinforced PP composites to make the dashboard carrier and chassis of their Alfa Romeo 4C model sports car.
- Europe will be the leading automotive composite market both in terms of cost and volume. New initiatives are being taken by automotive companies to increase the usage of PCs. An all-wheel-drive concept with carbon fiber composites is being developed by German car manufacturers for their safari rally car.
- The European Union is amongst the world's biggest producers of motor vehicles, supported by the largest private investor in research and development with an approximate investment of €57.4 billion annually. The turnover generated by European Union's auto industry contributes to 7% of the overall GDP.

The future scope of PCs in automotive applications is promising due to the following reasons:

- The demand for lightweight and fuel-efficient vehicles is increasing.
- The growing shift towards electric vehicles increases the growth of the automotive composites market.
- Novel processing techniques (RTM, Out-of-autoclave method), the arrival of fast-curing resins in the market will enhance the bulk production of PCs required for the automotive sector.
- A wide range of applications of composites in automotive applications such as exteriors, interiors, powertrains, and chassis keep the demand always higher.

However, there are challenges for using PCs for automotive applications as given below:

- High processing and manufacturing cost of composites.
- Lack of technological advancement in developing economies is a restraint.
- Limitations for high-temperature applications.
- Higher cost of production for few polymer composites for specific applications.

Another important aspect of PCs in automotive is the global push for zero emissions by 2050 that increases the demand for the development and production of electric vehicles (EVs). In September 2020, California announced it will require all new passenger cars and trucks sold in the state to be emission-free by 2035. Similarly, the EU proposed its 2030 target for new car CO_2 emissions is tightened to 50% below 2021 levels, up from 37.5%. As a global contribution to reducing CO_2 emission, many countries give tax benefits to their citizens for purchasing new electric vehicles. All these initiatives will significantly increase the demand for polymer matrix composites as they are lightweight, fuel economical, and eco-friendly.

CONCLUSION

PCs can make lightweight, more fuel-efficient yet safer vehicles. Today's automotive manufacturers use PCs to make several components, such as roofs, seats, steering wheel, hatch, dashboard, interior and exterior panel, engine cover, *etc.* The idea of using PCs in automotives is weight savings and possible cost savings as transport industries are becoming increasingly competitive and customer-friendly. A significant reduction in vehicle weight causes less fuel consumption and so CO_2 emissions. Compared to conventional structural materials like steel, iron, and aluminum, PCs offer weight savings of up to 25% for GFRP and 40% for CFRP composites. The lighter-weight CFRP composites will find extensive use in lower-volume niche vehicles, mainly for performance reasons. Corrosion resistance and easy part consolidation are the other advantages of PCs. However, lack of knowledge about material responses to automotive environments, unavailability of suitable crash models, and inadequate recycling technologies pose restrictions on the extensive use of PCs in automotive applications. The high production cost of carbon fibers also limits CFRP composites for specific applications. Extensive research and development may solve the current issues with PCs. Repairability, crash integrity, long-term durability, and recyclability would be thrust areas of research. From the environmental safety point of view, developing bio-based resins such as biodegradable polyesters and bio-based polyols will provide automotive manufacturers the provision to use bio-based composites for parts design.

ACKNOWLEDGEMENT

Declared none.

REFERENCES

[1]   Garcés JM, Moll DJ, Bicerano J, Fibiger R, McLeod DG. Polymeric nanocomposites for automotive applications. Adv Mater 2000; 12(23): 1835-9.
[http://dx.doi.org/10.1002/1521-4095(200012)12:23<1835::AID-ADMA1835>3.0.CO;2-T]

[2]   Ravishankar B, Nayak SK, Kader MA. Hybrid composites for automotive applications – A review. J Reinf Plast Compos 2019; 38(18): 835-45.
[http://dx.doi.org/10.1177/0731684419849708]

[3]   Santhanakrishnan Balakrishnan V, Seidlitz H. Potential repair techniques for automotive composites: A review. Compos, Part B Eng 2018; 145: 28-38.
[http://dx.doi.org/10.1016/j.compositesb.2018.03.016]

[4]   Friedrich K, Almajid AA. Manufacturing aspects of advanced polymer composites for automotive applications. Appl Compos Mater 2013; 20(2): 107-28.
[http://dx.doi.org/10.1007/s10443-012-9258-7]

[5]   Maciel MM, Ribeiro S, Ribeiro C, *et al.* Relation between fiber orientation and mechanical properties of nano-engineered poly(vinylidene fluoride) electrospun composite fiber mats. Compos, Part B Eng 2018; 139: 146-54.
[http://dx.doi.org/10.1016/j.compositesb.2017.11.065]

[6]   Beardmore P, Johnson CF. The potential for composites in structural automotive applications. Compos Sci Technol 1986; 26(4): 251-81.
[http://dx.doi.org/10.1016/0266-3538(86)90002-3]

[7]   Mallick P. Thermoplastics and thermoplastic–matrix composites for lightweight automotive structures. Materials, design and manufacturing for lightweight vehicles. Elsevier 2021; pp. 187-228.
[http://dx.doi.org/10.1016/B978-0-12-818712-8.00005-7]

[8]   Chaudhary V, Ahmad F. A review on plant fiber reinforced thermoset polymers for structural and frictional composites. Polym Test 2020; 91: 106792.
[http://dx.doi.org/10.1016/j.polymertesting.2020.106792]

[9]   Hovorun T. P., Berladir K. V., Pererva V. I., Rudenko S. G., Martynov A. I. Modern materials for automotive industry. Journal of Engineering Sciences. 2017; 4(2): F8-F18.
[http://dx.doi.org/10.21272/jes.2017.4(2).f8]

[10]   Das TK, Ghosh P, Das NC. Preparation, development, outcomes, and application versatility of carbon fiber-based polymer composites: a review. Adv Compos Hybrid Mater 2019; 2(2): 214-33.
[http://dx.doi.org/10.1007/s42114-018-0072-z]

[11]   Meng F, McKechnie J, Turner T, Wong KH, Pickering SJ. Environmental aspects of use of recycled carbon fiber composites in automotive applications. Environ Sci Technol 2017; 51(21): 12727-36.
[http://dx.doi.org/10.1021/acs.est.7b04069] [PMID: 29017318]

[12]   Kolesov YI, Kudryavtsev MY, Mikhailenko NY. Types and compositions of glass for production of continuous glass fiber. Glass Ceram 2001; 58(5/6): 197-202.
[http://dx.doi.org/10.1023/A:1012386814248]

[13]   Fuchs E, Field F, Roth R, Kirchain R. Strategic materials selection in the automobile body: Economic opportunities for polymer composite design. Compos Sci Technol 2008; 68(9): 1989-2002.
[http://dx.doi.org/10.1016/j.compscitech.2008.01.015]

[14]   Delmonte J. Metal/polymer composites. Springer 2013.

[15]   Zhang W, Xu J. Advanced lightweight materials for Automobiles: A review. Mater Des 2022; 221: 110994.
[http://dx.doi.org/10.1016/j.matdes.2022.110994]

[16]   Pruez J, Shoukry S, William G W., Shoukry M. Lightweight composite materials for heavy duty vehicles. West Virginia University 2013.
[http://dx.doi.org/10.2172/1116021]

[17]   Tripathi G. Application and future of composite materials: a review. International Journal of Composite and Constituent Materials 2017; 3(2): 1-4.

[18]   Gupta G, Kumar A, Tyagi R, Kumar S. Application and future of composite materials: A review. Int J

Innov Res Sci Eng Technol 2016; 5(5): 6907-11.

[19]   Mohammed L, Ansari M. N. M., Pua G, Jawaid M, Islam M S. A review on natural fiber reinforced polymer composite and its applications. International Journal of Polymer Science 2015.
[http://dx.doi.org/10.1155/2015/243947]

[20]   Fang H, Bai Y, Liu W, Qi Y, Wang J. Connections and structural applications of fibre reinforced polymer composites for civil infrastructure in aggressive environments. Compos, Part B Eng 2019; 164: 129-43.
[http://dx.doi.org/10.1016/j.compositesb.2018.11.047]

[21]   Kmetty Á, Bárány T, Karger-Kocsis J. Self-reinforced polymeric materials: A review. Prog Polym Sci 2010; 35(10): 1288-310.
[http://dx.doi.org/10.1016/j.progpolymsci.2010.07.002]

[22]   Henning F, Ernst H, Brüssel R. LFTs for automotive applications. Reinf Plast 2005; 49(2): 24-33.
[http://dx.doi.org/10.1016/S0034-3617(05)00546-1]

[23]   Kuyzin GS, Schlotterbeck DG, Kent GM. Low-density srim-non-glass reinforced polymer composites for automotive interior applications. Journal of Coated Fabrics 1990; 19(4): 211-29.
[http://dx.doi.org/10.1177/152808379001900403]

[24]   Chavhan GR, Wankhade LN. Experimental analysis of E-glass fiber/epoxy composite-material leaf spring used in automotive. Mater Today Proc 2020; 26: 373-7.
[http://dx.doi.org/10.1016/j.matpr.2019.12.058]

[25]   Sidhu, J., G. Lathkar, and S. Sharma, Design of epoxy based resin composites for automotive applications: a case study on IC engine valve guide. Journal of The Institution of Engineers (India): Series C, 2019. 100(2): p. 283-288.
[http://dx.doi.org/10.1007/s40032-017-0439-x]

[26]   Alavudeen A, Thiruchitrambalam M, Athijayamani A. Clutch plate using woven hybrid composite materials. Mater Res Innov 2011; 15(4): 229-34.
[http://dx.doi.org/10.1179/143307511X13018917925676]

[27]   Ishak NM, Sivakumar D, Mansor MR. The application of TRIZ on natural fibre metal laminate to reduce the weight of the car front hood. J Braz Soc Mech Sci Eng 2018; 40(2): 105.
[http://dx.doi.org/10.1007/s40430-018-1039-2]

[28]   Zulueta K, Burgoa A, Martínez I. Effects of hygrothermal aging on the thermomechanical properties of a carbon fiber reinforced epoxy sheet molding compound: An experimental research. J Appl Polym Sci 2021; 138(11): 50009.
[http://dx.doi.org/10.1002/app.50009]

[29]   Wang B, Ma S, Yan S, Zhu J. Readily recyclable carbon fiber reinforced composites based on degradable thermosets: a review. Green Chem 2019; 21(21): 5781-96.
[http://dx.doi.org/10.1039/C9GC01760G]

[30]   Savage G. Development of penetration resistance in the survival cell of a Formula 1 racing car. Eng Fail Anal 2010; 17(1): 116-27.
[http://dx.doi.org/10.1016/j.engfailanal.2009.04.015]

[31]   Sapuan SM, Suddin N, Maleque MA. A critical review of polymer-based composite automotive bumper systems. Polym Polymer Compos 2002; 10(8): 627-36.
[http://dx.doi.org/10.1177/096739110201000806]

[32]   Alves C, Ferrão PMC, Silva AJ, *et al.* Ecodesign of automotive components making use of natural jute fiber composites. J Clean Prod 2010; 18(4): 313-27.
[http://dx.doi.org/10.1016/j.jclepro.2009.10.022]

[33]   Ferreira F, Pinheiro I, de Souza S, Mei L, Lona L. Polymer composites reinforced with natural fibers and nanocellulose in the automotive industry: a short review. Journal of Composites Science 2019; 3(2): 51.

[http://dx.doi.org/10.3390/jcs3020051]

[34]   Bismarck A, Baltazar-Y-Jimenez A, Sarikakis K. -Jimenez, and K. Sarikakis, Green composites as panacea? Socio-economic aspects of green materials. Environ Dev Sustain 2006; 8(3): 445-63.
[http://dx.doi.org/10.1007/s10668-005-8506-5]

[35]   Malhotra SK, Goda K, Sreekala MS. Part one introduction to polymer composites. Polym Compos 2012; 1: 1-2.

[36]   Tanner D, Dhingra AK, Pigliacampi JJ. Aramid fiber composites for general engineering. J Miner Met Mater Soc 1986; 38(3): 21-5.
[http://dx.doi.org/10.1007/BF03257889]

[37]   Jagadale V, Taware S. A Review on Fatigue Analysis of Composite Steering Column. Engineering 2015.

[38]   Abdullah OG, Aziz SB, Rasheed MA. Effect of silicon powder on the optical characterization of Poly(methyl methacrylate) polymer composites. J Mater Sci Mater Electron 2017; 28(5): 4513-20.
[http://dx.doi.org/10.1007/s10854-016-6086-9]

[39]   Chawla KK. Composite materials: science and engineering. Springer Science & Business Media 2012.
[http://dx.doi.org/10.1007/978-0-387-74365-3]

[40]   McConnell VP. SMC has plenty of road to run in automotive applications. Reinf Plast 2007; 51(1): 20-5.
[http://dx.doi.org/10.1016/S0034-3617(07)70027-9]

[41]   Barbaz Isfahani R, Taherzadeh-Fard A, Haghighat Naeini E *et al.* Low velocity impact testing of glass fiber reinforced thermoplastic composites for the automotive industry. 2019.

[42]   Bilisik K, Kaya G, Ozdemir H, Korkmaz M, Erdogan G. Applications of glass fibers in 3D preform composites. Advances in Glass Science and Technology 2018; p. 207.
[http://dx.doi.org/10.5772/intechopen.73293]

[43]   Elsevier R. Reinforced Plastics–the voice of the composites industry worldwide. Global Automotive Composites Market Report 2021: Market is Projected to Grow from $5.4 Billion in 2020 to $9.3 Billion by 2025. 2021. Available From: https://www.prnewswire.com/news-releases/globa--automotive-composites-market-report-2021-market-is-projected-to-grow-from-5-4-b-llion-in-2020-to-9-3-billion-by-2025- -301213991.html

[44]   Witten E, Kraus T, Kühnel M. Composites Market Report 2015-Market developments, trends, outlook and challenges. AVK Industry Association for Reinforced Plastics, Carbon Composites eV 2015.

[45]   Meng F, Pickering S, McKechnie J. An environmental comparison of carbon fibre composite waste end-of-life options. Proceedings of the SAMPE Europe Conference 2018.

[46]   Holmes M. High volume composites for the automotive challenge. Reinf Plast 2017; 61(5): 294-8.
[http://dx.doi.org/10.1016/j.repl.2017.03.005]

[47]   Wan Y, Takahashi J. Development of carbon fiber-reinforced thermoplastics for mass-produced automotive applications in Japan. Journal of Composites Science 2021; 5(3): 86.
[http://dx.doi.org/10.3390/jcs5030086]

[48]   Ahmad H, Markina A A, Porotnikov M V, Ahmad F. A review of carbon fiber materials in automotive industry. Conf Ser: Mater Sci Eng 2020; 971.

[49]   Manson JA. HIVOCOMP: large-scale use of carbon composites in the automotive industry. Reinf Plast 2012; 56(6): 44-6.
[http://dx.doi.org/10.1016/S0034-3617(12)70150-9]

[50]   Gardiner, G. The markets: Automotive (2021). Reinforcements/Materials/Glass Fibers/Markets/Automotive 2020; Available from: https://www.compositesworld.com/articles/the-markets-automotive

[51]   Panthapulakkal S, Sain M. Injection-molded short hemp fiber/glass fiber-reinforced polypropylene hybrid composites—Mechanical, water absorption and thermal properties. J Appl Polym Sci 2007; 103(4): 2432-41.
[http://dx.doi.org/10.1002/app.25486]

[52]   Ramesh M, Bhoopathi R, Deepa C, Sasikala G. Experimental investigation on morphological, physical and shear properties of hybrid composite laminates reinforced with flax and carbon fibers. Journal of the Chinese Advanced Materials Society 2018; 6(4): 640-54.
[http://dx.doi.org/10.1080/22243682.2018.1534609]

[53]   Abhemanyu P. C., Prassanth E, Navin Kumar T, Vidhyasagar R, Prakash Marimuthu K, Pramod R. Characterization of natural fiber reinforced polymer composites. AIP Conference Proceedings. AIP Publishing LLC 2019.

[54]   Parameswaranpillai J, Joseph G, Shinu KP, Jose S, Salim NV, Hameed N. Development of hybrid composites for automotive applications: effect of addition of SEBS on the morphology, mechanical, viscoelastic, crystallization and thermal degradation properties of PP/PS–xGnP composites. RSC Advances 2015; 5(33): 25634-41.
[http://dx.doi.org/10.1039/C4RA16637J]

[55]   De Sciarra FM, Russo P. Experimental characterization, predictive mechanical and thermal modeling of nanostructures and their polymer composites. William Andrew 2018.

[56]   Nair A, Joseph R. Eco-friendly bio-composites using natural rubber (NR) matrices and natural fiber reinforcements. Chemistry, manufacture and applications of natural rubber. Elsevier 2014; pp. 249-83.
[http://dx.doi.org/10.1533/9780857096913.2.249]

[57]   Karus M. Use of natural fibres in composites for German automotive production from 1999 to 2005. Hemp: Industrial production and uses 2013; 187-94.

[58]   Natural Fiber Composites Market By Type (Wood, Non-Wood), By Manufacturing process (Injection Molding, Compression Molding, Pultrusion), By Application (Automotive, Electronics, Sporting Goods, Construction) Forecasts To 2027. ; Available from: https://www.reportsanddata.com/report-detail/natural-fiber-composites-market

[59]   Biocomposites: 350,000 t production of wood and natural fibre composites in the European Union in 2012, Renewable Carbon News.

[60]   Chand, N. and M. Fahim, Tribology of natural fiber polymer composites. tribology of natural fiber polymer composites, 2008: p. 1-205.
[http://dx.doi.org/10.1533/9781845695057]

[61]   Senthilkumar K, Saba N, Rajini N, *et al.* Mechanical properties evaluation of sisal fibre reinforced polymer composites: A review. Constr Build Mater 2018; 174: 713-29.
[http://dx.doi.org/10.1016/j.conbuildmat.2018.04.143]

[62]   Saxena M, Pappu A, Haque R, Sharma A. Sisal fiber based polymer composites and their applications. Cellulose fibers: Bio-and nano-polymer composites 2011; 589-659.
[http://dx.doi.org/10.1007/978-3-642-17370-7_22]

[63]   Shahzad A. Hemp fiber and its composites – a review. J Compos Mater 2012; 46(8): 973-86.
[http://dx.doi.org/10.1177/0021998311413623]

[64]   Ochi S. Mechanical properties of kenaf fibers and kenaf/PLA composites. Mech Mater 2008; 40(4-5): 446-52.
[http://dx.doi.org/10.1016/j.mechmat.2007.10.006]

[65]   Huang K, Ngoc Tran L Q, Kureemun U, Sze Teo W, Pueh Lee H. Vibroacoustic behavior and noise control of flax fiber-reinforced polypropylene composites. J Nat Fibers 2018; 729-43.

[66]   Wang Y, Wu H, Zhang C, *et al.* Acoustic characteristics parameters of polyurethane/rice husk composites. Polym Compos 2019; 40(7): 2653-61.

[http://dx.doi.org/10.1002/pc.25060]

[67]   Rajak D, Pagar D, Menezes P, Linul E. Fiber-reinforced polymer composites: Manufacturing, properties, and applications. Polymers (Basel) 2019; 11(10): 1667.
        [http://dx.doi.org/10.3390/polym11101667] [PMID: 31614875]

[68]   Huda MS, Drzal L.T., Ray D, Mohanty A.K, Mishra M. 7 - Natural-fiber composites in the automotive sector. Properties and performance of natural-fibre composites. Pickering KL, Ed. Woodhead Publishing 2008; pp. 221-68.
        [http://dx.doi.org/10.1533/9781845694593.2.221]

[69]   Alarifi IM, Movva V, Rahimi-Gorji M, Asmatulu R. Performance analysis of impact-damaged laminate composite structures for quality assurance. J Braz Soc Mech Sci Eng 2019; 41(8): 345.
        [http://dx.doi.org/10.1007/s40430-019-1841-5]

[70]   Djordjevic, B.B. Nondestructive test technology for the composites. in The 10th International Conference of the Slovenian Society for non-destructive testing, Ljubljana. 2009.

[71]   Jeong H. Effects of voids on the mechanical strength and ultrasonic attenuation of laminated composites. J Compos Mater 1997; 31(3): 276-92.
        [http://dx.doi.org/10.1177/002199839703100303]

[72]   V., O. Stankevych, and I. Kuz, Application of wavelet transforms for the analysis of acoustic-emission signals accompanying fracture processes in materials (a survey). Mater Sci 2018; 54(2): 139-53.
        [http://dx.doi.org/10.1007/s11003-018-0168-1]

[73]   Mouritz AP. Ultrasonic and interlaminar properties of highly porous composites. J Compos Mater 2000; 34(3): 218-39.
        [http://dx.doi.org/10.1177/002199830003400303]

[74]   David, K. Nondestructive inspection of composite structures: methods and practice. in 17th world conference on nondestructive testing, Shanghai. 2008. Citeseer.

[75]   Ghiorse SR. A comparison of void measurement methods for carbon/epoxy composites. Army Lab Command Watertown MA Material Technology Lab 1991.

[76]   Djordjevic B, Reis H. Sensors for materials characterization, processing, and manufacturing. ASNT Topics on NDE 1998; p. 1.

[77]   Kim YK, Chalivendra V. Natural fibre composites (NFCs) for construction and automotive industries. Handbook of Natural Fibres. Elsevier 2020; pp. 469-98.
        [http://dx.doi.org/10.1016/B978-0-12-818782-1.00014-6]

[78]   Strong AB. Fundamentals of composites manufacturing: Materials, methods and applications. Society of manufacturing engineers 2008.

[79]   Callister WD, Rethwisch DG. Materials science and engineering: an introduction. 2021.

[80]   Lattanzio RK, Tsang L, Canis B. Vehicle Fuel Economy and Greenhouse Gas Standards: Frequently Asked Questions. 2019. CRS Prepared for Members and Committees of Congress, Congressional Research.

[81]   Patel M, Pardhi B, Chopara S, Pal M. Lightweight composite materials for automotive-a review. Carbon 2018; 1(2500): 151.

[82]   Divya H, Naik LL, Yogesha B. Processing techniques of polymer matrix composites–A review. Int J Eng Res Gen Sci 2016; 4(3): 357-62.

[83]   Çakmakkaya M, Kunt M, Terzi O. Investigation of polymer matrix composites in automotive consoles. International Journal of Automotive Science And Technology 2019; 3(3): 51-6.
        [http://dx.doi.org/10.30939/ijastech..513332]

[84]   Khan L, Mehmood A. Cost-effective composites manufacturing processes for automotive applications. Ligh Comp Struct Trans. Elsevier 2016; pp. 93-119.

[http://dx.doi.org/10.1016/B978-1-78242-325-6.00005-0]

[85]   Qureshi J. A review of fibre reinforced polymer structures. Fibers (Basel) 2022; 10(3): 27.
       [http://dx.doi.org/10.3390/fib10030027]

[86]   Global Automotive Composites Market by Fiber Type (Glass, Carbon, Natural), Resin Type
       (Thermoset, Thermoplastics), Manufacturing Process (Compression, Injection, RTM), Applications
       (Exterior, Interior), Vehicle Type and Region - Forecast to 2025. 2021; Available from:
       https://www.prnewswire.com/news-releases/global-automotive-composites-market-r-
       port-2021-market-is-projected-to-grow-from-5-4-billion-in-2020-to-9-
       3-billion-by-2025--301213991.html

SUBJECT INDEX

A

Acids 10, 32, 33, 65, 72, 75, 76, 78, 79, 126, 130, 141, 148, 155, 178
  amino 79
  ascorbic 32, 33
  glycolic 78
  hyaluronic 65
  hydrochloric 75
  lactic 78, 79
  nitric 10
  nucleic 148
  perfluorosulfonic 126, 130
  phosphoric 75
  polylactic 72, 76, 178
  sulfuric 75
  sulphuric 10
AEM in electrochemical processes 128, 142
Alkaline fuel cell (AFC) 126
Antibacterial activity 34
Antimicrobial 29, 31, 34, 35, 41, 42
  action 29
  activity 31, 34, 35, 42
  agents 29, 31, 35
  macromolecules 41
Applications 34, 43, 58, 59, 67, 78, 79, 89
  biomedical 58, 59, 67, 78, 79
  electrochemical 89
  electronic 43
  encapsulation 34
Automobiles 58, 59, 97, 98, 166
  electric 97
Automotive industries 166, 167, 168, 170, 171, 172, 175, 176, 177, 185, 187, 188

B

Biosensors 37, 39, 40, 44, 154
  antigen-detecting 37
Bolometric sensor 155
Bone tissue engineering (BTE) 72, 74, 75, 76, 80

Bovine serum albumin (BSA) 154
BTE scaffold 78

C

Cathode-electrolyte framework 124
Ceramic(s) 64, 66, 67, 185
  conventional 67
  matrix composites (CMC) 64, 66, 67, 185
Chemical vapour deposition (CVD) 33, 149, 179
Composite-based additive manufacturing (CBAM) 71
Conducting polymers (CPs) 43, 88, 98, 137, 146, 147, 148, 153, 154, 155
Conductor 139
  electronic 139
Corrosion 3, 6, 10, 65, 169, 171, 190
  chemical resistance properties 171
  problems 10
  protection 3
  resistance 6, 65, 169, 190
CVD method 149
Cycles 3, 128
  electrodialysis-related 128
Cytotoxicity 81

D

Defects, irregular-shaped bone 76
Deformations 38, 154, 180
  mechanical 38, 154
Deformed electron clouds 151
Degradation 11, 22, 35, 41, 59, 77, 132
  mechanical 59
Deposition 33, 43, 88
  physical vapour 33
  process 43
  vapor 88
Devices 35, 38, 40, 43, 61, 87, 90, 92, 99, 100, 146, 153
  analytical 40